CALCULUS AND GRAPHS

THE MACMILLAN COMPANY
NEW YORK · BOSTON · CHICAGO · DALLAS
ATLANTA · SAN FRANCISCO

MACMILLAN & CO., Limited
LONDON · BOMBAY · CALCUTTA
MELBOURNE

THE MACMILLAN CO. OF CANADA, Ltd.
TORONTO

CALCULUS AND GRAPHS

SIMPLIFIED FOR A FIRST BRIEF COURSE

BY

L. M. PASSANO

ASSOCIATE PROFESSOR OF MATHEMATICS
IN THE MASSACHUSETTS INSTITUTE OF TECHNOLOGY
AUTHOR OF "PLANE AND SPHERICAL TRIGONOMETRY"

New York
THE MACMILLAN COMPANY
1921

PREFACE

The importance of the natural sciences is so generally recognized as to need no emphatic statement. Nor is it necessary to point out the dependence of the sciences upon the study and knowledge of mathematics. This dependence is closer and more direct in the case of calculus than in the case of any other branch of mathematics, unless, perhaps, we except elementary algebra and trigonometry. It is primarily for the purpose of making the elements of the calculus directly and familiarly available to students of physics, chemistry and other sciences that the present book is written. At the same time it is hoped that the book will be found well adapted to the use of those who wish an elementary knowledge of calculus for its cultural value.

No knowledge of Analytic Geometry is assumed on the part of students using the present text. On the other hand the idea of coördinate axes and their use in the graphical representation and study of simple algebraic and transcendental functions is introduced in the first chapter and used continually throughout the work. The student becomes familiar with the fundamental ideas of analytic geometry, learns to use both algebraic and geometric methods in the study of functions, and becomes acquainted with the forms and equations of simple curves without definition of those curves or detailed study of their properties. The student thus acquires all the knowledge of analytic geometry necessary to an understanding of the elements of calculus; and assuming on his part a knowledge of elementary algebra and trigonometry, the calculus is made available for a first college course.

Such a course, the writer believes, is not only more useful

v

but also more interesting and simple than a first course in analytic geometry. It is not intended, however, that the present work should replace the study of analytic geometry, the importance of which should not be underrated, but that the study of the latter subject should simply be deferred until the more immediately important subject, the calculus, has been acquired.

The author has striven to present the subject in as simple a way as possible. The aim has been to make the student understand the subject; not to write a book that would satisfy meticulous mathematical pedantry. In so doing the author may have in places sacrificed logical detail to simplicity of presentation, but never, he hopes, accuracy of statement. In the opinion of the writer a too rigidly logical proof with its paraphernalia of subscripted Greek letters is out of place in an elementary first course in calculus, for the reason that the student never understands such a proof. Or if by arduous effort he does grasp its meaning it is at the expense—in time and labor—of other things that are more important and far more useful.

The author wishes to thank a number of his colleagues for criticism and suggestions in the writing of the present work. He is especially indebted to Professor C. L. E. Moore and Professor D. P. Bartlett for their criticism of the manuscript and for improvements which they suggested.

L. M. PASSANO.

Massachusetts Institute of Technology
  Cambridge, Mass.

CONTENTS

CONTENTS

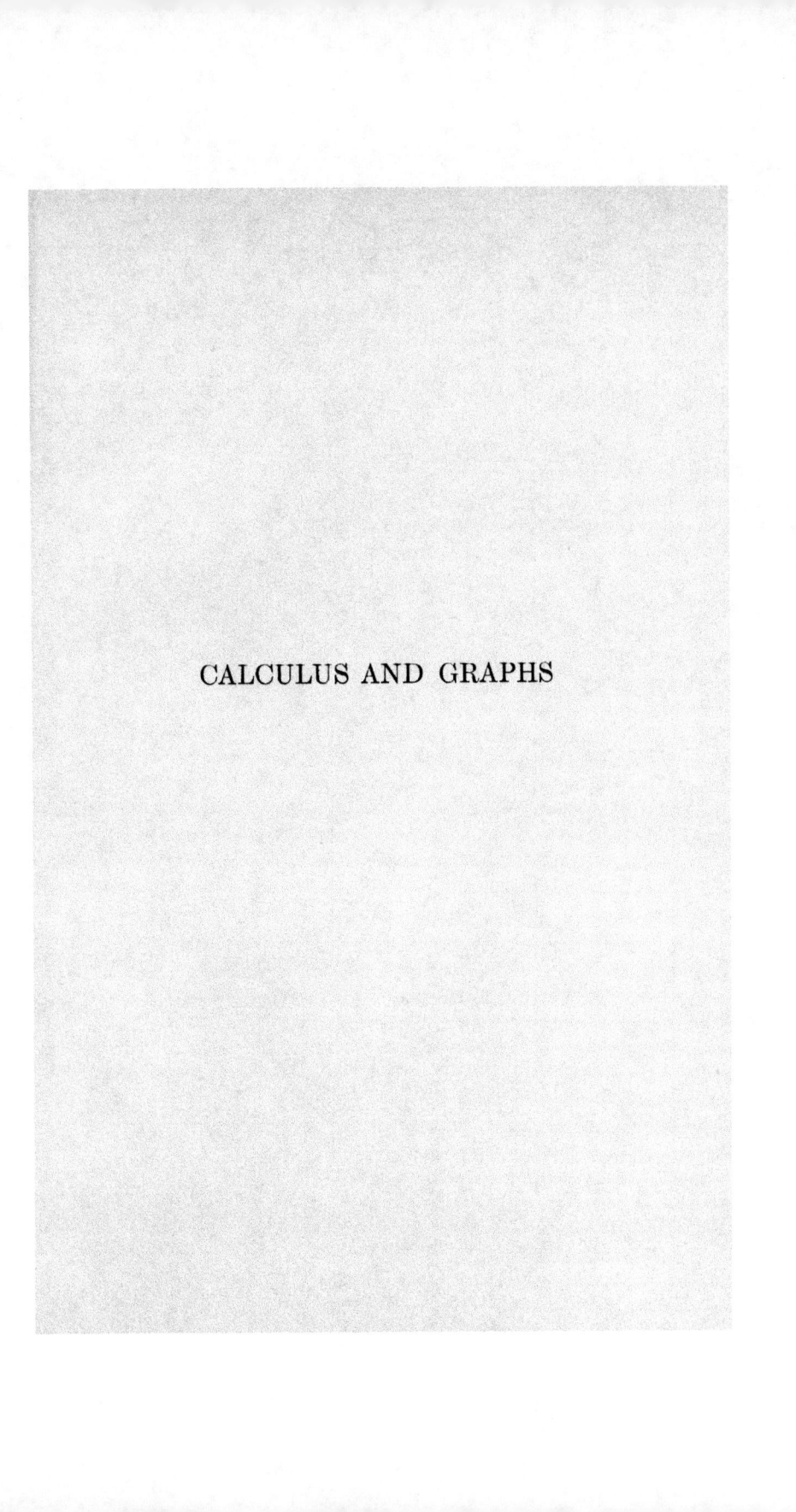

CALCULUS AND GRAPHS

CALCULUS AND GRAPHS

CHAPTER I

VARIABLES, FUNCTIONS, AND GRAPHS

1. Variables and Functions. In dealing with numbers
one is brought to consider two well marked kinds: constants
and variables. Their names suggest their nature. A
constant number, or quantity is one which, under the conditions of the problem in which it occurs, does not change
in value; a variable is one which takes different values.
Whether the values of either are known or unknown makes
no difference. Of two numbers, a and x, say, we may not
know the single actual value of a, and may know that the
x must have the definite range of values from 2 to 6. The
number a would still be a constant and the number x a
variable.

In many, or most, of our problems it happens also that
we have to deal with *two* variables, and any number of
known or unknown constants; and with two variables so
related that the value of one depends upon the value of
the other. For example, the room in which we are sitting,
the doors and windows being shut, will have a certain
temperature which varies. We can change the temperature by increasing or decreasing the amount of steam in
the radiator. The amount of steam we can regulate in any
way we please, but the temperature we can regulate only
by means of the steam, upon which it depends. We express this mathematically by saying that the temperature,
a dependent variable, is a *function* of the steam, an inde-

1

pendent variable, and we write the relation thus, using y to represent temperature and x steam,

$$y = f(x), \text{ or } y = F(x) \quad y = \phi(x) \text{ etc.;}$$

read, y equals f of x, ϕ of x, etc., or, in general, y equals a function of x.

Many simple examples of functions suggest themselves. For example, $A = S^2$ expresses the fact that the area of a square (A) is a particular function of its side (S); $c = 2\pi r$, expresses that the circumference of a circle (c) is a particular function of its radius (r). By giving values to s in the one case or to r in the other one can find the corresponding values of A or c. Thus, when the side of the square is 2 feet the area is 4 square feet; when s is 12 feet, $A = 144$ sq. ft.; when r is 7 inches c is $14\pi = 44$ inches.

EXAMPLES

1. Express the volume of a cube as a function of the edge. Find the value of the volume when the edge is 2 feet, 7 feet, 3.1 centimeters.

2. Express the volume and surface of a sphere each as a function of the radius; the volume as a function of the surface.

3. The height of a rectangular pyramid is three times the side of its base. Express the volume of the pyramid as a function of the side of the base; as a function of the height.

4. The illumination, I, from a source of light varies directly as the candle power, c, and inversely as the square of the distance, d. If the distance is 10 feet express the illumination as a function of the candle power. If the candle power is 50, express the illumination as a function of the distance. Which gives the greater illumination .5 candle power at 10 feet or 50 candle power at 100 feet?

Find the values of the following functions when $x = 1$ and when $x = 4$; also the amount by which the function increases or decreases as x changes in value from 1 to 4.

5. $y = x^2 + 2x$ 10. $y = 2^x$

6. $s = 3\sqrt{x}$ 11. $y = x - 3x^2$

7. $u = 2x + \dfrac{1}{2x}$

8. $y = \log_{10} x$

9. $y = \tan x$

12. $y = x - x^3$

13. $y = -5x + \dfrac{1}{x}$

14. $y = -x^2 + x$

2. A System of Rectangular Coördinates. Let us draw in a plane two straight lines at right angles to each other, (Fig. 1).

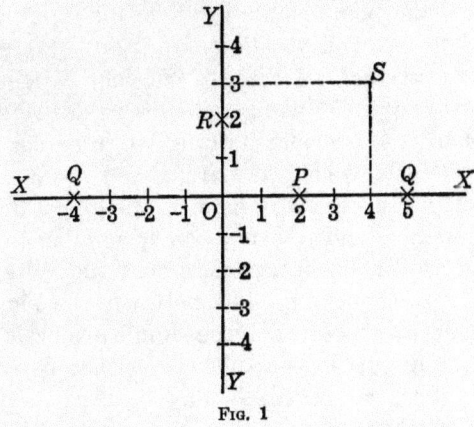

FIG. 1

Let us call the point, O, in which the lines intersect, the origin, the lines themselves the axes, of the system of rectangular coördinates. To distinguish the two axes one, the line XOX is called the axis of x or the axis of abscissas; the other YOY, the axis of y or the axis of ordinates. Suppose we wish to distinguish the points on OX from each other. To do so we associate with each point a symbol, called a number. This might be done in any way, but for the sake of system and order we will lay off any convenient unit of length along XOX and attach to the points thus got to the right of O the series of numbers $1, 2, 3, 4 \ldots \ldots$; to those to the left of O the series $-1, -2, -3, -4, \ldots \ldots \ldots$ Thus, using fractional and incommensurable numbers as well as integers, there corresponds to every point on XOX a certain number and to every number a point on XOX. In the same way to every point on YOY corresponds a number, and to every number a point on YOY. But given the number 2 we should not know whether the point P were meant or the point R (Fig. 1). To remove this ambiguity we say, when the point P is meant, the point whose abscissa, or x, is 2; when R is meant, the point whose ordinate, or y, is 2; written respectively $x = 2$ and $y = 2$.

Suppose next that we wish to consider a point S, not lying on either axis. Obviously its abscissa, distance along or parallel to OX is 4 and its ordinate, distance along or parallel to OY is 3, which we write $x = 4, y = 3$ *, or more briefly (4, 3). Thus to every point in the plane corresponds a pair of numbers and to every pair of numbers a point in the plane. It will be noticed also that the system of co-ordinates is a connecting link between the simple geometrical concept, a point, and the simple algebraic concept, a number, and thus in general between the two branches of mathematics; so that each may be used as a help in studying the other.

3. Functions and Graphs: Plotting Graphs. Consider y a simple function of x; say, $y = 2x + 3$. We can give the independent variable x any values we please, and compute the corresponding values of the function y, as shown in the following table:

x	0	1	2	3	-1	-2	-3	etc.
y	3	5	7	9	1	-1	-3	etc.

* Obviously any other letters might be used. Thus OT might be the axis of abscissas and OS the axis of ordinates. The point would then be written $t = 4, s = 3$.

We thus have seven pairs of numbers, values of x and y, and, as we have seen, each such pair of numbers corresponds to a point in a system of coördinates. If we mark these points, plot them as it is called, we shall have a picture, a graphic representation, of the functional relation connecting. the two variable numbers x and y as shown in Fig. 2.

If we take values of x at small intervals we can get as many points as we please as near together as we please, the graph appearing more and more nearly as a continuous line. We thus speak of y as a continuous function of x; a concept that will be more strictly defined later on. (See Art. 5.) We can now study y, the function of x, by using the graph and the methods of Geometry or by using the

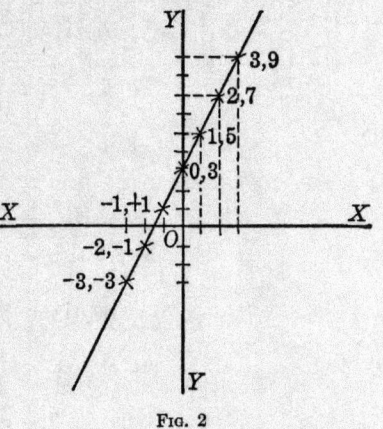

Fig. 2

equation and the methods of Algebra. In many cases, indeed, we use both. Note most carefully that if a point lies on the graph its coördinates must satisfy the equation of the graph; if a pair of values of the coördinates satisfy the equation the corresponding point must lie on the graph of the function. Also if a point lies at the same time on two (or more) graphs, its coördinates will satisfy the two (or more) equations of the graphs. If we wish to find values of the variables (x and y, say) that satisfy two equations we may sketch the graphs of the equations and measure the abscissas and ordinates of their points of intersection; or, if we wish to find the points of intersection of two graphs we can solve simultaneously the equations of the graphs. Thus the graphs of $y = 2x - 3$ and $x + y = 4$ intersect in the point $x = \frac{7}{3}$, $y = \frac{5}{3}$.

Let us consider other simple functions.

Example 1. $y = 2x^2 - 8$.   See Fig. 3.

x	0	1	2	3	-1	-2	-3	etc.
y	-8	-6	0	10	-6	0	10	etc.

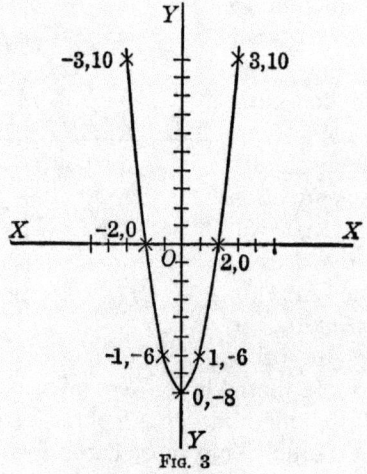

Fig. 3

Example 2. $y = \sin x$.   See Fig. 4.

In sketching this graph it is convenient, since x must be measured along OX, to express the angle in circular measure.

x	0	$\frac{\pi}{6}$	$\frac{\pi}{3}$	$\frac{\pi}{2}$	$\frac{2\pi}{3}$	$\frac{5\pi}{6}$	π	$\frac{7\pi}{6}$	$\frac{4\pi}{3}$	$\frac{3\pi}{2}$	$\frac{5\pi}{3}$	$\frac{11\pi}{6}$	2π
y	0	.50	.87	1.00	.87	.50	0	$-.50$	$-.87$	-1.00	$-.87$	$-.50$	0

We need not give values to x greater than 2π because the trigonometric functions are periodic; that is, the same value of the function will occur at regular intervals of 2π. Note that, as a matter of convenience, different units are used on the two axes.

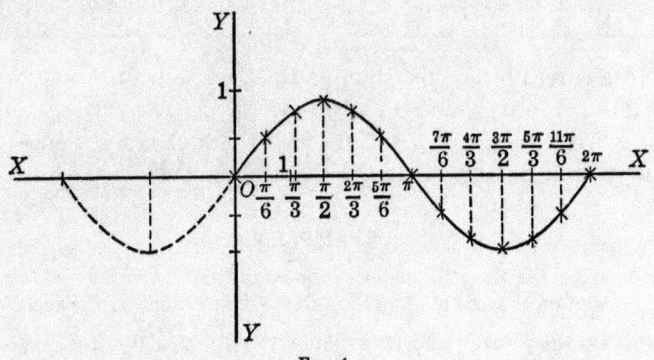

FIG. 4

The graph extends indefinitely to right and left in repetition of the part drawn, as is indicated by the dotted portion in Fig. 4.

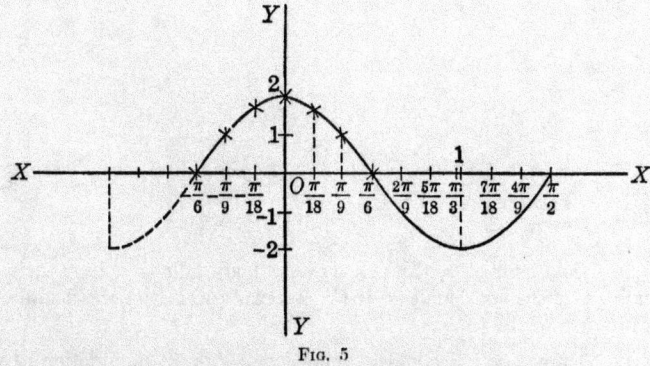

FIG. 5

Example 3. $y = 2 \cos 3 x$. See Fig. 5.

x	0	$\dfrac{\pi}{18}$	$\dfrac{\pi}{9}$	$\dfrac{\pi}{6}$	$\dfrac{2\pi}{9}$	$\dfrac{5\pi}{18}$	$\dfrac{\pi}{3}$	$\dfrac{7\pi}{18}$	$\dfrac{4\pi}{9}$	$\dfrac{\pi}{2}$	$-\dfrac{\pi}{18}$	$-\dfrac{\pi}{9}$	$-\dfrac{\pi}{6}$
$3x$	0	$\dfrac{\pi}{6}$	$\dfrac{\pi}{3}$	$\dfrac{\pi}{2}$	$\dfrac{2\pi}{3}$	$\dfrac{5\pi}{6}$	π	$\dfrac{7\pi}{6}$	$\dfrac{4\pi}{3}$	$\dfrac{3\pi}{2}$	$-\dfrac{\pi}{6}$	$-\dfrac{\pi}{3}$	$-\dfrac{\pi}{2}$
y	2	1.74	1	0	-1	-1.74	-2	-1.74	-1	0	1.74	1	0

Note carefully in this example that, while we use values of $3x$ to compute values of y, we *plot the points by using values of x with values of y.* Note also that for convenience we use a smaller unit along OY than along OX.

EXAMPLES

1. What are the coördinates of the origin? What is the abscissa of any point on the axis of y? The ordinate of any point on the axis of x?

2. Do the points $(1, 3)$ $(1, 4)$ $(0, -1)$ $(3, -2)$ lie on the graph of the function $y = 5x - 1$? Do they lie on the graph of $y = 2x^2 + 1$? Why?

3. What points are represented by the equation $x = 3$? By the equation $y = -2$? By $x = 0$? By $y = 0$?

Plot the graphs of the following functions:

4. $2y = x + 4$ 9. $y = x - x^2$

5. $y = 3x$ 10. $y = \sin\dfrac{x}{2}$

6. $y = \dfrac{x}{4} - 3$ 11. $y = 3\cos 2x$

7. $y = (x - 1)^2$ 12. $y = \tan x$

8. $y = \dfrac{x^3}{3}$ 13. $y = \sec x$

14. The distance, s, through which a body falls in a time, t, is expressed by the equation $s = 16\,t^2$. Express graphically the distance as a function of the time.

15. Express graphically the volume of a cube as a function of the edge.

16. Draw a graph showing the area of a circle as a function of the radius.

17. A particle moves so that its distance from the axis of y is given by $x = 2\,t$, its distance from the axis of x by $y = t - 2$, where t represents the time, in seconds, the particle has moved. Plot the path of the body during the first 10 seconds. From what point does the particle start? At what point is it at the end of the tenth second?

Find the points of intersection of the following pairs of graphs:

18. $y = 2\,x - 3$
$x = y$

19. $2\,x + 3\,y = 7$
$x - y = 4$

20. $x + 2\,y = 2$
$3\,x - 4\,y = 1$

21. $y = x$
$\dfrac{x^2}{9} + \dfrac{y^2}{16} = 1$

22. $y = x + 1$
$x^2 + y^2 = 16$

23. $y = x$
$y = 4\,x^3$

24. $y = 2$
$y = 3 \sin x$

25. $y = 1$
$y = \cos \dfrac{x}{2}$

26. $y = \sqrt{3}$
$y = \tan x$

27. Two particles move in straight lines, passing through an origin 0, according to the laws $s = 3\,t - 5$ and $s = t + 4$ respectively. At what time (t) will the two particles be at equal distances (s) from the origin? What is the distance?

Two particles move in a straight line passing through an origin 0, according to the laws $s = t + 4$ and $s = t^2 - 5$.

28. When will the two particles be at equal distances from 0? What is the distance?

29. When will the first particle be twice as far from 0 as the second? What is the distance of each?

30. When will the second particle be twice as far from 0 as the first? What is the distance of each?

4. Sketching Graphs: Critical or Peculiar Points. In many of our problems it happens that we are interested not so much in the exact figure of the graph representing a function as in its general features; whether, for instance, it is a closed or open curve; what are its limits; the position of critical or peculiar points. Such information can usually

be got, without accurately plotting the graph, by *sketching* it with especial reference to its peculiarities. The following examples illustrate some of the methods employed for the ends mentioned.

Example 1. $y + 2 = 3 (x - 1)^2.$

Put $y + 2 = y', \quad x - 1 = x'$

The equation then becomes $y' = 3 x'^2.$

It is obvious that numerically equal positive and negative values of x' give equal values of y', so that the curve is symmetrical with respect to OY'.* Also, x' occurring as a square only, y' can not be negative. The least value of y', zero, is got when x' is numerically least $(= 0)$. The curve can, therefore, be sketched readily without the labor of obtaining numerical points. Fig. 6.

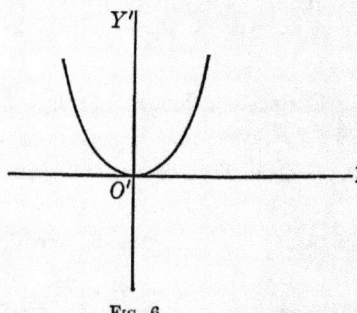

FIG. 6

This is not, however, the graph of the function originally given but of the simplified function $y' = 3 x'^2$. But, by the assumptions we see that $y = y' - 2$ and $x = x' + 1$; that is, the distance (y) from the true axis of x is 2 less than the distance from OX' and the distance (x) from the true axis of y is 1 more than the distance from OY'. We therefore draw OX parallel to and 2 units above OX' and OY parallel to and 1 unit to the left of OY'. The axis of symmetry of the curve is now seen to be $x = 1$, and the vertex of the curve $x = 1, y = -2$, as shown in Fig. 7.

* In general it is worth while to note that if the equation contains even powers only of $x(y)$ it is symmetrical with respect to $OY(OX)$.

Example 2. $y = \sin\left(x + \dfrac{\pi}{4}\right).$

Put $x + \dfrac{\pi}{4} = x'$ so that $y = \sin x'$.

We can simplify
the sketching of y
$= \sin x'$ by using
our knowledge of
the trigonometric
functions. The
greatest value of
the sine of an an-
gle is 1; the least
value, -1; the
least numerical
value, 0. There-
fore, put $\sin x' = 0$.
$x' = \sin^{-1} 0$, and
we have

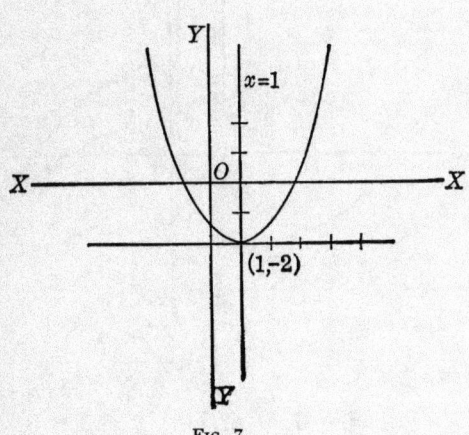

Fig. 7

x'	0	π	2π	etc.
y	0	0	0	etc.

the points where the graph crosses OX'.

Put $\sin x' = 1, x' = \sin^{-1} 1$ Put $\sin x' = -1, x' = \sin^{-1}(-1)$

x'	$\dfrac{\pi}{2}$	$\dfrac{5\pi}{2}$	etc.
y	1	1	etc.

x'	$\dfrac{3\pi}{2}$	$\dfrac{7\pi}{2}$	etc.
y	-1	-1	etc.

the highest points on the   the lowest points on the
graph.   graph.

We can now sketch the graph by means of these critical points and fix the original axes as in Example 1 above. See Fig. 8 and compare Example 2, Art. 3.

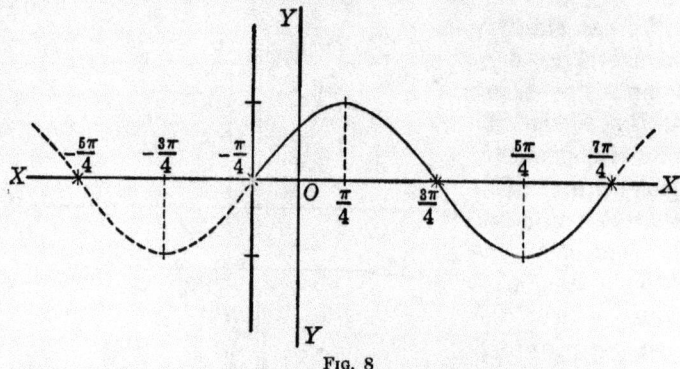

Fig. 8

Example 3. $y = x^3 + 2\,x^2 = x^2\,(x + 2)$
     Put $x^2 = 0$ and $x + 2 = 0$
       or $x = 0$ $x = -\,2$

thus obtaining critical points. Then by taking values of x less than and greater than the critical values proceed, as shown in the following table, to determine the peculiarities of the graph and to sketch it. Fig. 9.

x	-3	-2	-1	0	1
y	$-$	0	$+$	0	$+$

The only points actually found are $(-2, 0)$ and $(0, 0)$. Since y is negative to the left of $x = -2$ the curve comes up from below the axis of x, which it crosses at $x = -2$; remains above OX between $x = -2$ and $x = 0$, since in this interval y is positive; touches the axis of x at $x = 0$, and extends above OX thenceforth because y remains positive for values of x greater than $x = 0$. It must be noted that the graph is not accurately plotted but merely sketched. We do not know, for instance, the highest point on the arch of the curve; that is, the greatest value of the function lying between the values when $x = -2$ and $x = 0$. Later in our work we shall learn how to determine such values.

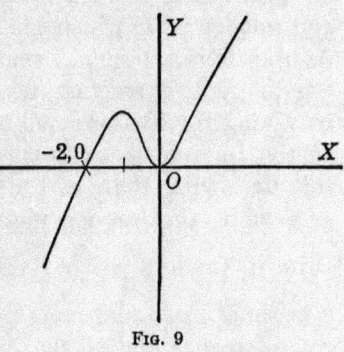

Fig. 9

Example 4. $xy - 2x + 2y + 3 = 0$.

In this case y may be regarded as a function of x or x as a function of y. Solving the equation for y and x respectively we have

$$y = \frac{2x - 3}{x + 2}, \qquad x = \frac{3 + 2y}{2 - y}.$$

Using the former and putting

$$2x - 3 = 0 \qquad x + 2 = 0$$
$$x = \tfrac{3}{2} \qquad x = -2$$

we have critical points. As in Example 3, $x = \tfrac{3}{2}$ gives $y = 0$, but $x = -2$ gives $y = -\tfrac{7}{0}$. What does this mean? Zero does not mean " nothing." Mathematics does not concern itself with things that mean nothing. On the contrary the science prides itself on the fact that every concept and symbol used has a very definite and exact meaning, so that once the " language " is mastered there can be no doubt as to the meaning of every word and sentence.

The symbols $\frac{7}{0}$ or, in general, $\frac{a}{0}$, mean simply that a (or 7) is being divided by a number which grows smaller and smaller * and which can be made as small as we please. As the divisor becomes smaller and smaller the value of the quotient, or fraction, becomes greater and greater; and by taking the divisor small enough we can make the value of the fraction as great as we please. We may express this by saying that, as the denominator approaches zero as a limit, the fraction increases without limit, and may write it, briefly, $\frac{a}{0} = \infty$. Thus in the above example, as x becomes more and more nearly equal to -2, $x + 2$ becomes smaller and smaller, and by taking x as near -2 as we please (on either side) we can make $x + 2$ as small as we please. We express this briefly by saying that when x equals minus two, x plus 2 equals zero and the fraction equals infinity, and write

$$x = -2,\ x + 2 = 0,\ \text{and}\ y = -\frac{7}{0} = \infty.$$

We do not know, however, whether y is plus infinity or

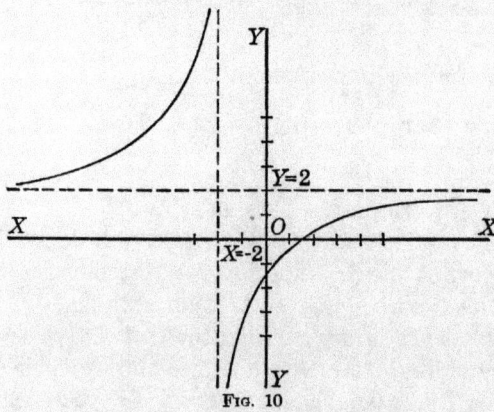

FIG. 10

* Numerically; that is, without regard to the algebraic sign.

minus infinity, since we do not yet know whether $x + 2$ is plus zero or minus zero; that is, whether $x + 2$ is an exceedingly small positive or an exceedingly small negative number. We will proceed to sketch our graph, however, by marking (Fig. 10) the point $(\frac{3}{2}, 0)$ where it crosses OX, and, where $x = -2$, by drawing a line parallel to OY. We will also tabulate our values in order, finding values, or the sign, of y for values of x less than and greater than the critical values.

x	-3	-2	0	$\frac{3}{2}$	2
y	$+9$	∞	$-\frac{3}{2}$	0	$\frac{1}{4}$

Since y is positive to the left of $x = -2$ we see that the ordinate is positive and very large just to the left of $x = -2$; expressed, $y = +\infty$. Since y is negative between $x = -2$ and $x = \frac{3}{2}$ we see that the ordinate is negative and very large just to the right of $x = -2$; expressed, $y = -\infty$. We might now proceed to sketch the graph, but before doing so let us consider the other form of the equation,

$$x = \frac{3 + 2y}{2 - y}.$$

Put $\quad 3 + 2y = 0, \quad y = -\frac{3}{2} \quad$ and we find $x = 0$, a point already discussed.

Put $\quad 2 - y = 0, \; y = 2 \quad$ and we find $x = \infty$. We, therefore, draw a line $(y = 2)$ parallel to OX, and tabulate our values.

y	-2	$-\frac{3}{2}$	0	2	3
x	$-\frac{1}{4}$	0	$\frac{3}{2}$	∞	-9

Reasoning as above we find $x = -\infty$ for positions just above the line $y = 2$, and $x = +\infty$ just below the line $y = 2$. The graph can now be completely sketched as shown. The lines $x = -2$ and $y = 2$ are called asymptotes of the curve. They meet the curve at a distance greater than any number one may please to assign however great; that is, at infinity.

EXAMPLES

Sketch the graphs of the following functions:

1. $y - 3 = (x - 1)^2$

2. $y + 4 = 2(2x - 3)^2$

3. $x = (y - 2)^2$

4. $x = (y + 3)^2$

5. $y = x^3 + x$

6. $y = x^3 - x$

7. $x = y^3 - 2y^2$

8. $x = y^3 + 3y^2$

9. $y = \sin(x - \frac{\pi}{3})$

10. $y = \sin(x + 2\pi)$

11. $y + 2 = 2\sin x$

12. $y - 3 = \sin 2x$

13. $y = \cos(x + \frac{1}{3})$

14. $y = \cos(x - \frac{1}{2})$

15. $y = \tan(x - \frac{\pi}{2})$

16. $y = \tan(x + \frac{\pi}{3})$

17. $xy - x + 2y = 1$

18. $xy - x - 2y + 1 = 0$

19. $2xy - x + 6y + 2 = 0$

20. $2xy + x + 6y = 2$

21. $xy - x - by + a = 0$

22. $xy + x + by = a$

23. $s = t^3 - a^2t$

24. $s = t^3 + bt^2$

25. The product of the pressure (p) and volume (v) of a perfect gas is constant and equal in a particular case, to 4. Express the volume as a function of the pressure and sketch the graph. What is the volume when the pressure equals 1000? What happens when the pressure increases without limit? Does v ever become zero? Does the volume ever vanish?

26. A particle moves in a straight line so that $s = t^3 - 3t^2$, where s is the distance in feet from an origin on the line and t is the time in seconds. Sketch the graph of the function. By means of your graph

determine at what time the particle will be at the origin on the straight line. Does it move to the left or to the right of the origin? How long?

27. A particle moves in a straight line so that $s = t^2 - 3t + 2$, s and t being distance and time as in Example 26. Sketch the graph of the function. By means of your graph determine from what point on the straight line the particle starts to move. Does the particle ever reach the origin on the straight line? When? Does it move to the left or to the right of the origin? How long?

28. A particle moves so that its distance from OY is always $x = t$; its distance from OX is $y = t^2 + 2t$, where t is the time in seconds. Sketch the path of the particle. Where does the particle start to move? Does it ever move backward in its path? Why?

INCREMENTS. DIFFERENTIATION AND THE DE-RIVATIVE. THEOREMS AND FORMULAS OF DIFFERENTIATION FOR ALGEBRAIC AND TRIGO-NOMETRIC FUNCTIONS

5. Increments: Continuous Functions. In many problems it is of greater interest and importance to know how and by how much a variable is changing than to know the actual value of the variable itself. In the case of the independent variable this is not so, because by the very nature of an independent variable we can make the change anything we please. But if we make a change in the value of an independent variable, x, what change is caused in the value of a variable, y, which depends upon x; in the function of x which y represents? Let us consider a simple case, $y = x^2$. Suppose x has the value 2 (or 4) and that we make changes in its value, give it various increments, as they are called, where the increment may be either positive or negative. The resulting values and changes in value of y are shown in the following table, where the symbol Δx (delta x) means the increment given to x and Δy the increment caused in y by Δx.

x	y	Δx	Δy	$\dfrac{\Delta y}{\Delta x}$	x	y	Δx	Δy	$\dfrac{\Delta y}{\Delta x}$
2	4	...	...	...	4	16	...	...	...
3	9	1	5	5	5	25	1	9	9
2.1	4.41	.1	.41	4.1	4.1	16.81	.1	.81	8.1
2.01	4.0401	.01	.0401	4.01	4.01	16.0801	.01	.0801	8.01

The tables show clearly that the increment of the function (Δy) depends both upon the value of the independent variable (x) and its increment (Δx); and also show that Δy grows smaller as Δx does. It would seem from the table that by making Δx small enough we could make Δy as small as we please. When this is the case, when the increment of the function can be made smaller than any number we may assign, however small, by making the increment of the independent variable small enough—as it is usually expressed: if the increment of the function approaches zero as a limit when the increment of the independent variable approaches zero as a limit; written $\lim_{\Delta x = 0} \Delta y = 0$ — in such case y is called a continuous function of x.*

The last column of the tables gives the value of the ratio of the increment of the function to the increment of the independent variable, which ratio also is seen to depend upon both x and Δx. We note also that, when $x = 2$, the value of $\frac{\Delta y}{\Delta x}$ as Δx grows smaller becomes more and more nearly equal to 4; when $x = 4$, the value of $\frac{\Delta y}{\Delta x}$ becomes more nearly equal to 8 as Δx grows smaller. We might ask what will be the value of Δy and more particularly of $\frac{\Delta y}{\Delta x}$ when Δx has become smaller than any

---

* That $\lim_{\Delta x = 0} \Delta y$ is not necessarily zero may be indicated by the simple function

$y = \frac{1}{x}$.

As x grows nearer to $-.1$ (Δx decreases) Δy increases rapidly.

x	y	Δx	Δy
$-.1$	-10		
$.1$	$+10$	$.2$	20
$.01$	100	$.11$	110
$.001$	1000	$.101$	1010
$.0001$	10000	$.1001$	10010

number we may please to assign. That is are we justified
in saying that for the function $y = x^2$ when

$$x = 2, \quad \underset{\Delta x = 0}{\text{limit}} \ \frac{\Delta y}{\Delta x} = 4, \text{ and}$$

$$x = 4, \quad \underset{\Delta x = 0}{\text{limit}} \ \frac{\Delta y}{\Delta x} = 8?$$

To answer this question let us try to generalize our
process. Let $\Delta x = h$, any number. Then the new value
of the function will be $y + \Delta y = (2 + h)^2$ and

$$\Delta y = (y + \Delta y) - y = (2 + h)^2 - (2)^2$$

which may be written

$$\Delta y = 4 + 4h + h^2 - 4 = 4h + h^2$$

and

$$\frac{\Delta y}{\Delta x} = 4 + h.$$

Therefore, passing to the limit, since $h = \Delta x$,

$$\underset{\Delta x = 0}{\text{limit}} \ \frac{\Delta y}{\Delta x} = \underset{h = 0}{\text{limit}} \ (4 + h) = 4$$

Thus, we see, our conclusion was correct; we " guessed
right." In the same way we could show that for $x = 4$
the result would be 8, and by the same process we could
find the result for any other numerical value of x. But
let us generalize again and use any value of x whatever.
Thus

$$y = x^2, \ y + \Delta y = (x + h)^2, \text{ and}$$

$$\Delta y = (x + h)^2 - x^2 = 2hx + h^2,$$

$$\frac{\Delta y}{\Delta x} = 2x + h,$$

$$\underset{\Delta x = 0}{\text{limit}} \ \frac{\Delta y}{\Delta x} = \underset{h = 0}{\text{limit}} \ (2x + h) = 2x.$$

Now, by simply putting any numerical value for x in our result, we can find the value of $\displaystyle\lim_{\Delta x = 0} \frac{\Delta y}{\Delta x}$ for the function $y = x^2$ for any value of x whatever.

The above process is fundamentally important and can be applied to any of the functions with which we deal. It must be thoroughly mastered.

Example 1. Let $y = \dfrac{1}{x}$, and $\Delta x = h$, then

$$y + \Delta y = \frac{1}{x + h}$$

$$\Delta y = \frac{1}{x + h} - \frac{1}{x} = \frac{-h}{x^2 + hx}$$

$$\frac{\Delta y}{\Delta x} = -\frac{1}{x^2 + hx}$$

$$\lim_{\Delta x = 0} \frac{\Delta y}{\Delta x} = \lim_{h = 0}\left(-\frac{1}{x^2 + hx}\right) = -\frac{1}{x^2}$$

which when $x = 1, 2, 3$, etc., takes the values -1, $-\frac{1}{4}$, $-\frac{1}{9}$, etc.

Example 2. $y = x^2 + x, \Delta x = h,$

$$y + \Delta y = (x + h)^2 + (x + h)$$

$$\Delta y = (x + h)^2 + (x + h) - x^2 - x = 2xh + h^2 + h$$

$$\frac{\Delta y}{\Delta x} = 2x + h + 1$$

$$\lim_{\Delta x = 0} \frac{\Delta y}{\Delta x} = 2x + 1.$$

EXAMPLES

In the following functions find the limit of the ratio of the increment of the function to the increment of the independent variable as the increments approach zero as a limit, showing each step of the process.

1. $y = 2x^2 - 3$ 2. $y = \dfrac{x^2}{4} + x$ 3. $y = 3x^2 - 2x$

4. $s = t^3 + 1$ 5. $u = v^3 + \dfrac{1}{v}$ 6. $s = t^2 - \dfrac{1}{t^2}$

7. $xy = x + 1$ 8. $x^2y = x^3 + x^{-4}$ 9. $y = \sqrt{x}$

10. $y = \dfrac{x+1}{x-2}$ 11. $y = \dfrac{1-x}{3+x}$ 12. $pv = a$

6. The Derivative: Differentiation. The expression, $\lim\limits_{\Delta x = 0} \dfrac{\Delta y}{\Delta x}$, found in Art. 5 is called a derivative; the process of finding a derivative is called differentiation. Thus, *y being a continuous function of x the derivative of y with respect to x is the limit of the ratio of the increment of y to the increment of x as the increments approach zero as a limit.*

The notation used in Art. 5 is exact but awkward. The student having now learned the exact meaning of *derivative* we may use a simpler notation, thus:

$$\lim_{\Delta x = 0} \frac{\Delta y}{\Delta x} = \frac{dy}{dx} \tag{1}$$

where dy and dx are not used separately, and $\dfrac{dy}{dx}$ (read, the derivative of y with respect to x; or, dy over dx) is not as yet to be treated as a fraction, though we shall treat it so later and in many cases use dy and dx advantageously as separate quantities.

The fundamental process (Art. 5) being, however, too cumbersome for frequent use we proceed to generalize once more, in order to obtain processes simpler and more direct. Indeed Mathematics might be called a "lazy man's delight," because it is continually devising methods to make our work shorter and less laborious. At the same time, whenever a new one of these intellectual "tools" is invented, it is found to be wonderfully adapted to the solution of new problems. Among these intellectual tools the simple derivative, $\dfrac{dy}{dx}$, is perhaps the most wonderful.*

We shall now proceed to simplify the process of finding a derivative, considering first algebraic functions. The operations of Algebra are addition (including subtraction), multiplication, division (which may be reduced to that of multiplication by the use of negative exponents), raising to powers, and extraction of roots (reducible to raising to powers by the use of fractional exponents). We need only consider then differentiation as applied to functions involving addition, multiplication and raising to powers.

7. The Derivative of a Sum. Let $y = u + v + w + \ldots$ where $u, v, w \ldots$ are functions † of an independent variable x. Give x an increment Δx, thus $u, v, w \ldots$ will receive increments $\Delta u, \Delta v, \Delta w \ldots$, and y an increment Δy. Then

$$y + \Delta y = u + \Delta u + v + \Delta v + w + \Delta w + \ldots$$

$$\Delta y = (u + \Delta u + v + \Delta v + w + \Delta w + \ldots) - (u + v + w + \ldots)$$

or $\quad \Delta y = \Delta u + \Delta v + \Delta w + \ldots$

$$\frac{\Delta y}{\Delta x} = \frac{\Delta u}{\Delta x} + \frac{\Delta v}{\Delta x} + \frac{\Delta w}{\Delta x} + \ldots$$

and $\quad \underset{\Delta x = 0}{\text{limit}} \ \dfrac{\Delta y}{\Delta x} = \underset{\Delta x = 0}{\text{limit}} \left(\dfrac{\Delta u}{\Delta x} + \dfrac{\Delta v}{\Delta x} + \dfrac{\Delta w}{\Delta x} + \ldots \right)$

$$= \underset{\Delta x = 0}{\text{limit}} \ \frac{\Delta u}{\Delta x} + \underset{\Delta x = 0}{\text{limit}} \ \frac{\Delta v}{\Delta x} +$$

$$\underset{\Delta x = 0}{\text{limit}} \ \frac{\Delta w}{\Delta x} + \ldots$$

* The weight of the evidence seems to give to Sir Isaac Newton (1642–1727 A.D.) the honor of the invention of the calculus.

† Continuous functions. The word continuous will be dropped henceforth.

or, as we now write,

$$\frac{dy}{dx} = \frac{d(u + v + w + \ldots)}{dx} = \frac{du}{dx} + \frac{dv}{dx} + \frac{dw}{dx} + \ldots \quad (2)$$

which is memorized briefly as " The derivative of a sum is equal to the sum of the derivatives." *

8. The Derivative of a Product. Let $y = uv$ where u and v are functions of x. Proceeding as above

$$y + \Delta y = (u + \Delta u)(v + \Delta v)$$

$$\Delta y = (u + \Delta u)(v + \Delta v) - uv$$

$$= v \Delta u + u \Delta v + \Delta u \Delta v$$

$$\frac{\Delta y}{\Delta x} = v \frac{\Delta u}{\Delta x} + u \frac{\Delta v}{\Delta x} + \Delta u \frac{\Delta v}{\Delta x}$$

$$\lim_{\Delta x = 0} \frac{\Delta y}{\Delta x} = \lim_{\Delta x = 0} \left(v \frac{\Delta u}{\Delta x} + u \frac{\Delta v}{\Delta x} + \Delta u \frac{\Delta v}{\Delta x} \right)$$

$$= \lim_{\Delta x = 0} v \frac{\Delta u}{\Delta x} + \lim_{\Delta x = 0} u \frac{\Delta v}{\Delta x} +$$

$$\lim_{\Delta x = 0} \Delta u \frac{\Delta v}{\Delta x}$$

or

$$\frac{dy}{dx} = \frac{d(uv)}{dx} = v \frac{du}{dx} + u \frac{dv}{dx} \qquad (3)$$

memorized briefly as, " the derivative of the product of two factors equals the second factor times the derivative of the first, *plus* the first factor times the derivative of the second."

---

* This is not self-evident, nor does experience lead us to think it true. The square of a sum is *not* the sum of the squares; the square root of a sum is *not* the sum of the square roots; the sine of a sum is *not* the sum of the sines, etc.

Note in (3) that $\displaystyle \lim_{\Delta x = 0} \Delta u \frac{\Delta v}{\Delta x} = 0$ because Δu approaches zero as Δx does.

9. The Derivative of a Power. Let $y = u^n$ where u is a function of x. As above

$$y + \Delta y = (u + \Delta u)^n$$

$$\Delta y = (u + \Delta u)^n - u^n$$

$$= u^n + n u^{n-1} \Delta u + \frac{n(n-1)}{1.2} u^{n-2} \overline{\Delta u}^2 + \ldots$$

$$\ldots\ldots + \overline{\Delta u}^n - u^n$$

$$\frac{\Delta y}{\Delta x} = n u^{n-1} \frac{\Delta u}{\Delta x} + \frac{n(n-1)}{1.2} u^{n-2} \Delta u \cdot \frac{\Delta u}{\Delta x} + \ldots$$

$$+ \overline{\Delta u}^{n-1} \frac{\Delta u}{\Delta x}$$

$$\lim_{\Delta x = 0} \frac{\Delta y}{\Delta x} = \lim_{\Delta x = 0} n u^{n-1} \frac{\Delta u}{\Delta x} +$$

$$\lim_{\Delta x = 0} \frac{n(n-1)}{1.2} u^{n-2} \Delta u \frac{\Delta u}{\Delta x} + \ldots$$

or

$$\frac{dy}{dx} = \frac{du^n}{dx} = n u^{n-1} \frac{du}{dx} \qquad (4)$$

In particular if $u = x$

$$\frac{dx^n}{dx} = n x^{n-1} \qquad (5)$$

Theorems (4) and (5) have been proved for positive integral values only of n. They are true for fractional and negative exponents, but the proof is omitted.

We now have the following eight theorems, or formulas, of differentiation.

10. Collected Formulas of Differentiation.

i. $\dfrac{d(u + v + w + \ldots)}{dx} = \dfrac{du}{dx} + \dfrac{dv}{dx} + \dfrac{dw}{dx} + \ldots$

ii. $\dfrac{d(uv)}{dx} = v\dfrac{du}{dx} + u\dfrac{dv}{dx}$

iii. $\dfrac{du^n}{dx} = nu^{n-1}\dfrac{du}{dx}$

iv. $\dfrac{dx^n}{dx} = nx^{n-1}$

v. $\dfrac{dx}{dx} = 1$ (6)

vi. $\dfrac{dcu}{dx} = c\dfrac{du}{dx}$; c a constant.

vii. $\dfrac{dc}{dx} = 0$; c a constant.

viii. $\dfrac{d\dfrac{u}{v}}{dx} = \dfrac{v\dfrac{du}{dx} - u\dfrac{dv}{dx}}{v^2}$

The proofs of the last four being left to the student. To these we add,

i. $\dfrac{d\sin u}{dx} = \cos u\dfrac{du}{dx}.$ iii. $\dfrac{d\sec u}{dx} = \sec u \tan u\dfrac{du}{dx}$

ii. $\dfrac{d\tan u}{dx} = \sec^2 u\dfrac{du}{dx}$ iv. $\dfrac{d\cos u}{dx} = -\sin u\dfrac{du}{dx}$

v. $\dfrac{d\cot u}{dx} = -\csc^2 u\dfrac{du}{dx}$ (7)

vi. $\dfrac{d\csc u}{dx} = -\csc u \cot u\dfrac{du}{dx}$

To prove the formula $\dfrac{d \sin u}{dx} = \cos u \dfrac{du}{dx}$ we proceed as follows: Draw a circle (Fig. 11) with centre at 0 and with

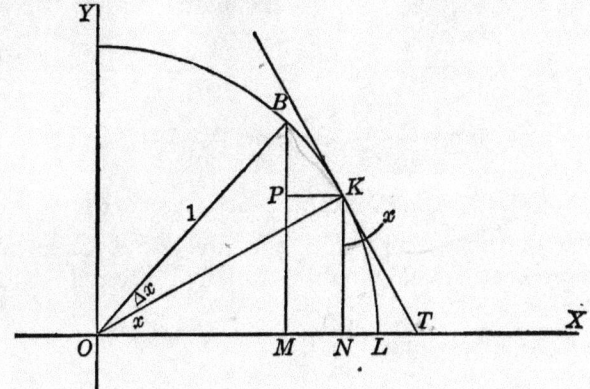

FIG. 11

radius unity (so taken for convenience). Let the angle $LOK = x$, measured in radians, and angle $LOB = x + \Delta x$. Draw KN and BM perpendicular to OX and KP perpendicular to $BM;$ also KT tangent to the circle at K. Obviously the angle $NKT = x$, $NK = \sin x$ and $MB = \sin (x + \Delta x)$. Also arc $KB = \Delta x$ since the angles are expressed in circular measure, and, by trigonometry, arc $KB = \Delta x$ times radius $OK = \Delta x$. We thus have

$$PB = \Delta \sin x$$

$$\frac{PB}{\text{arc } KB} = \frac{\Delta \sin x}{\Delta x}$$

$$\underset{KB=0}{\text{limit}} \frac{PB}{\text{arc } KB} = \underset{\Delta x=0}{\text{limit}} \frac{\Delta \sin x}{\Delta x} = \frac{d \sin x}{dx} \qquad \text{(a)}$$

Also

$$\frac{PB}{\text{chord } KB} = \cos PBK$$

$$\underset{KB=0}{\text{limit}} \ \frac{PB}{\text{chd. } KB} = \text{limit} \cos PBK$$

$$= \cos \text{ limit } PBK$$

$$= \cos NKT = \cos x \qquad \text{(b)}$$

But

$$\text{limit} \frac{PB}{\text{arc } KB} = \text{limit} \frac{PB}{\text{chd. } KB}$$

since, when the point B approaches the point K as its limiting position, chord KB and arc KB become more and more nearly equal. As it is usually expressed

$$\text{limit} \frac{\text{chord } KB}{\text{arc } KB} = 1 *$$

Therefore, from (a) and (b) we have

$$\frac{d \sin x}{dx} = \cos x$$

We see again that, u being a continuous function of x,

$$\frac{\Delta \sin u}{\Delta x} = \frac{\Delta \sin u}{\Delta u} \cdot \frac{\Delta u}{\Delta x}$$

and

$$\underset{\Delta x=0}{\text{limit}} \frac{\Delta \sin u}{\Delta x} = \underset{\Delta u=0}{\text{limit}} \frac{\Delta \sin u}{\Delta u} \cdot \underset{\Delta x=0}{\text{limit}} \frac{\Delta u}{\Delta x}$$

or

$$\frac{d \sin u}{dx} = \cos u \frac{du}{dx}$$

To prove the formula

$$\frac{d \cos u}{dx} = - \sin u \frac{du}{dx}$$

* This statement is not proved but is taken as sufficiently obvious for our purpose.

we might proceed as in the case of the derivative of the sine, using Fig. 11, or as follows:

$$\frac{d \cos u}{dx} = \frac{d \sin \left(\frac{\pi}{2} - u\right)}{dx} = \cos \left(\frac{\pi}{2} - u\right) \frac{d\left(\frac{\pi}{2} - u\right)}{dx}$$

$$= \sin u \frac{d(-u)}{dx} = -\sin u \frac{du}{dx}$$

Also, since $\tan u = \dfrac{\sin u}{\cos u}$, $\sec u = \dfrac{1}{\cos u}$ etc., the remaining formulas of group (7) can be proved by using the derivatives of sine and cosine together with the formulas of group (6), Art. 10. These proofs are left as an exercise for the student.

By means of a few examples we will illustrate the use of the formulas of differentiation.

Example 1. $y = 2 x^3 + 3 x - 2 x^{1/2}$

$$\frac{dy}{dx} = \frac{d(2 x^3 + 3 x - 2 x^{1/2})}{dx}$$

which is the derivative of a sum. Therefore, by (6)—i

$$\frac{dy}{dx} = \frac{d \, 2 x^3}{dx} + \frac{d \, 3 x}{dx} - \frac{d \, 2 x^{1/2}}{dx}$$

and by (6)—vi

$$\frac{dy}{dx} = 2 \frac{dx^3}{dx} + 3 \frac{dx}{dx} - 2 \frac{dx^{1/2}}{dx}$$

whence by (6)—iv and v

$$\frac{dy}{dx} = 2.3 \, x^2 + 3.1 - 2.\tfrac{1}{2} \, x^{-1/2} = 6 \, x^2 + 3 - \frac{1}{x^{1/2}}$$

Example 2. $y = \dfrac{x^2 + 2}{x - 1}$

or $y = (x^2 + 2) \, (x - 1)^{-1}$

$$\frac{dy}{dx} = (x-1)^{-1}\frac{d(x^2+2)}{dx} + (x^2+2)\frac{d(x-1)^{-1}}{dx} \text{ by (6)—ii}$$

$$= (x-1)^{-1}\left\{\frac{dx^2}{dx} + \frac{d\,2}{dx}\right\} + (x^2+2)\left\{ - (x-1)^{-2}\right.$$

$$\left. \frac{d(x-1)}{dx}\right\}$$

by (6)—i   (6)—iii

$$= (x-1)^{-1}(2x+0) + (x^2+2)\left\{-(x-1)^{-2}\left[\frac{dx}{dx} - \frac{d1}{dx}\right]\right\}$$

by (6)—iv; vii   (6)—i

$$= (x-1)^{-1}\,2\,x - (x^2+2)\,(x-1)^{-2}\,(1-0) \text{ by (6)—v; vii}$$

$$\frac{dy}{dx} = \frac{2x}{x-1} - \frac{x^2+2}{(x-1)^2} = \frac{x^2-2x-2}{(x-1)^2}$$

Example 3. $y = 2\sin^2 3\,x$

$$\frac{dy}{dx} = 2\frac{d\sin^2 3\,x}{dx} = 2.2\sin 3\,x\,\frac{d\sin 3\,x}{dx}$$

(6)—vi   (6)—iii

$$= 4\sin 3\,x\,\cos 3\,x\,\frac{d\,3\,x}{dx} = 4\sin 3\,x\,\cos 3\,x\,.\,3$$

(7)—i   (6)—vi

$$= 12\sin 3\,x\,\cos 3\,x = 6\sin 6\,x$$

Example 4. $y = x^2\tan 2\,x$

$$\frac{dy}{dx} = \tan 2\,x\,\frac{dx^2}{dx} + x^2\frac{d\tan 2\,x}{dx} \qquad \text{(6)—ii}$$

$$= \tan 2\,x\,.\,2\,x + x^2\sec^2 2\,x\,\frac{d\,2\,x}{dx}$$

(6)—iv   (7)—ii

$$= 2\,x\,\tan 2\,x + x^2\,\sec^2 2\,x\,.\,2 \qquad (6)\text{—vi}$$

$$= 2\,x\,(\tan 2\,x + x\,\sec^2 2\,x)$$

Example 5. $\quad x^2 y^3 + 2\,xy = 7\,x^2$

$$\frac{d(x^2 y^3 + 2\,xy)}{dx} = \frac{d\,7\,x^2}{dx}$$

$$\frac{d(x^2 y^3)}{dx} + \frac{d(2\,xy)}{dx} = \frac{7\,dx^2}{dx}$$

$$(6)\text{—i} \qquad\qquad (6)\text{—vi}$$

$$y^3 \frac{dx^2}{dx} + x^2 \frac{dy^3}{dx} + 2\left\{ y\,\frac{dx}{dx} + x\,\frac{dy}{dx} \right\} = 7\,.\,2\,x$$

$$(6)\text{—ii and vi}$$

$$y^3\,.\,2\,x + x^2\,.\,3\,y^2\,\frac{dy}{dx} + 2\,y + 2\,x\,\frac{dy}{dx} = 14\,x$$

$$(6)\text{—iv; \quad iii; \quad v}$$

$$2\,xy^3 + 3\,x^2 y^2\,\frac{dy}{dx} + 2\,y + 2\,x\,\frac{dy}{dx} = 14\,x$$

which is an algebraic equation of the first degree in $\dfrac{dy}{dx}$. Whence

$$(3\,x^2 y^2 + 2\,x)\,\frac{dy}{dx} = 14\,x - 2\,y - 2\,xy^3$$

$$\frac{dy}{dx} = \frac{14\,x - 2\,y - 2\,xy^3}{2\,x + 3\,x^2 y^2}.$$

Note: In equations such as that in Example 5 the y depends upon x for its value and is, therefore, a function of x, but as the value of y in terms of x is not exactly expressed but merely implied, we say that y is an *implicit* function of x. When y is exactly expressed in terms of x we call y an *explicit* function.

EXAMPLES

In each of the following examples find the derivative of the function.

1. $3x^3 - 2x + 4$

2. $x^2 - x^4 + 3x$

3. $(x + 3)^3$

4. $(2 + x)^{-2}$

5. $(x^2 + 2)^3$

6. $(3 - x^3)^2$

7. $(x^3 + x)^{1/3}$

8. $(x - 2x^2)^{1/2}$

9. $\sin x \cos x$

10. $\sin x \tan x$

11. $\cos x + \sec 2x$

12. $\sin 2x - \tan 3x$

13. $\dfrac{1}{u+2} + \dfrac{1}{u-2}$

14. $\dfrac{2}{S} - \dfrac{3}{S+7}$

15. $\dfrac{5}{y(y^2+2)}$

16. $\dfrac{3}{y - y^4}$

17. $3x^{2.7}$

18. $5x^{-5.02}$

19. $\tfrac{1}{2} x^{3.14}$

20. $-2x^{-2^{1/4}}$

21. $x^2 + y^2 = 16$

22. $4x^2 + 5y^2 = 20$

23. $4x^2 - 5y^2 = 20$

24. $xy = 100$

25. $xy + x - 2y + 7 = 0$

26. $xy - 3x - y = 4$

27. $\sin^2 (2x + \tfrac{\pi}{3})$

28. $\cot^3 (x - \tfrac{\pi}{4})$

29. $\cos x \cdot \sec (2x + \tfrac{\pi}{6})$

30. $\sin x \cos 2x \tan 3x$

CHAPTER III

APPLIED MEANINGS AND USES OF THE DERIVATIVE. RATE OF CHANGE. VELOCITY. ACCELERATION

11. Rate of Change. Velocity. Acceleration. Suppose there is a quantity, u, which changes in value as time elapses, and suppose that a change (increment) of time Δt causes an increment, Δu, in u. Then $\dfrac{\Delta u}{\Delta t}$ will be the *average* rate of change of u during the time Δt. For example, suppose the temperature of the room has risen from 40° to 70° between 9:00 A. M. and 11:00 A. M. Then the temperature has risen at the average rate of 15° per hour or .25° per minute.

$$\frac{\Delta u}{\Delta t} = \frac{70-40}{11-9} = \frac{30}{2} = 15.$$

Can we say that at 10:00 o'clock the temperature was changing at that same rate? Obviously not, because at 9:55 o'clock the temperature may have risen to 80°, the heating steam may have then been turned off and the temperature, stationary, perhaps, for 5 minutes, have fallen between 10:00 and 11:00 A. M. from 80° to 70°. Thus the temperature at 10:00 A. M. may have been actually falling, instead of rising at the rate of .25° per minute. Again suppose we walk to school from home, a distance of 3 miles, in 45 minutes. Our average rate of walking, our average speed, the average rate of change of the distance, would be 3 miles $\div \frac{3}{4}$ hours = 4 miles per hour. Can we say we crossed Avenue A at 10th Street at 4 miles per

hour? Probably not. If the automobiles were numerous
our speed (rate of change of distance) was more likely to
be six or eight miles per hour. How then can we find the
rate of change of a quantity at a given instant, the instantaneous rate of change, and what is meant by instantaneous rate of change? Suppose Avenue A is 52.8
feet broad and we crossed it in 6 seconds. Our average
speed across the avenue was 6 miles per hour. But when
two-thirds the way across we may have speeded up to
dodge a car, so that the last 17.6 feet were covered in one
second. Our average speed over this distance was, therefore, 12 miles per hour. How can we find the speed at
which we mounted the curb on the far side of the avenue?
We see how it is done. By taking the distance and time
shorter and shorter we find the average speed through a
shorter and shorter distance and time, and by taking the
distance and time short enough we obtain the average
speed for a time and distance so short that we can call it
the instantaneous speed. Thus:

Instantaneous speed $=$ limit $\dfrac{\text{distance}}{\text{time}}$ as distance and
time approach zero.

In mathematical symbols

$$\text{Speed} = v = \lim_{\Delta t = 0} \frac{\Delta s}{\Delta t} = \frac{ds}{dt} \qquad (8)$$

and, in general, if u is any quantity which changes in value,
the instantaneous

$$\text{Rate of change of } u = \lim_{\Delta t = 0} \frac{\Delta u}{\Delta t} = \frac{du}{dt} \qquad (9)$$

Obviously, Δt being taken positive, if $\dfrac{\Delta u}{\Delta t}$ is $\begin{cases} \text{positive} \\ \text{negative} \end{cases}$

so also is Δu, and therefore u is $\left\{\begin{array}{c}\text{increasing}\\\text{decreasing}\end{array}\right\}$. But if

$\dfrac{\Delta u}{\Delta t}$ is always $\left\{\begin{array}{c}\text{positive}\\\text{negative}\end{array}\right\}$ then also must $\displaystyle\lim_{\Delta t=0}\dfrac{\Delta u}{\Delta t}=$

$\dfrac{du}{dt}$ be $\left\{\begin{array}{c}\text{positive}\\\text{negative}\end{array}\right\}$ and we can say that if $\dfrac{du}{dt}$ is $\left\{\begin{array}{c}\text{positive}\\\text{negative}\end{array}\right\}$

u must be an $\left\{\begin{array}{c}\text{increasing}\\\text{decreasing}\end{array}\right\}$ function.

It must be noted that speed is merely a special case of rate of change; that is, the rate of change of a distance. It occurs so frequently that a special name, speed, is given to it. Another rate of change, of frequent occurrence, to which a special name is given is the rate of change of the speed. The rate of change of speed is called acceleration. In symbols

$$\text{Acceleration} = f = \frac{dv}{dt} \qquad (10)$$

which may also be written

$$f = \frac{d\left(\dfrac{ds}{dt}\right)}{dt} = \frac{d^2s}{dt^2} \qquad (11)$$

In general $\dfrac{d^2u}{dx^2}$ is called the second derivative of u with respect to x, and means the derivative of the derivative.

12. Angular Velocity and Acceleration. Another important instance of rate of change is that which arises in the case of a rotating or revolving body. Suppose a wheel turns on its axis, making 100 revolutions per minute. (Fig. 12); then 100 revolutions per minute = 200π radians per minute is called the angular velocity. In this case the angular velocity is constant, but if the wheel is coming to rest because of the friction of the air and of its axle, the angular velocity will not be constant. Reasoning as in previous instances we may say that

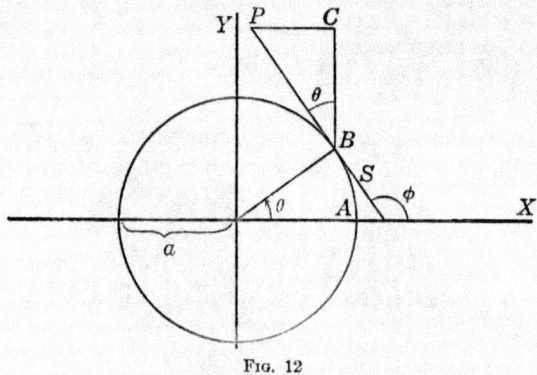

Fig. 12

$$\frac{\Delta \theta}{\Delta t} = \text{average angular velocity during the time } \Delta t$$

and

$$\frac{d \theta}{dt} = \underset{\Delta t = 0}{\text{limit}} \frac{\Delta \theta}{\Delta t} = \text{angular velocity at any instant} = \omega. (12)$$

Similarly

$$\frac{d\omega}{dt} = \frac{d\left(\dfrac{d \theta}{dt}\right)}{dt} = \frac{d^2 \theta}{dt^2} = \alpha = \text{the rate of change of the}$$

angular velocity; called angular acceleration   (13)

As we learned in Trigonometry, the angle θ being expressed in circular measure or radians, (see Fig. 12).

$$\theta = \frac{s}{a} \text{ or } s = a\theta$$

whence

$$\frac{ds}{dt} = a\frac{d\theta.}{dt} \tag{14}$$

where $\frac{ds}{dt}$ is the speed of a point on the rim of a wheel of radius a. The motion of the point is in the direction of the tangent to the curve at any instant.

13. Components of Velocity. At the point B (Fig. 12) let BP represent in length and direction the velocity of the point at any instant. Then, as we have learned in our study of Physics, BP can be resolved into two components, CP parallel to OX and BC parallel to OY.

From the figure

$$CP = BP \sin\theta \quad \text{and} \quad BC = BP \cos\theta$$

where CP and BC are the rates of change of distances x and y parallel to OX and OY respectively. We can, therefore, write

$$\frac{dx}{dt} = -\frac{ds}{dt}\sin\theta \quad \text{and} \quad \frac{dy}{dt} = \frac{ds}{dt}\cos\theta$$

or

$$v_x = -v\sin\theta \qquad v_y = v\cos\theta \tag{15}$$

It is obvious that

$$v_x^2 + v_y^2 = v^2(\sin^2\theta + \cos^2\theta) = v^2.$$

That is, the speed of a particle in its path equals the square root of the sum of the squares of its component speeds parallel to the axes of x and y.

Similarly (Fig. 13) if a particle is moving along a path in a system of rectangular coördinates, the distance from some initial point, A, being represented by S and the

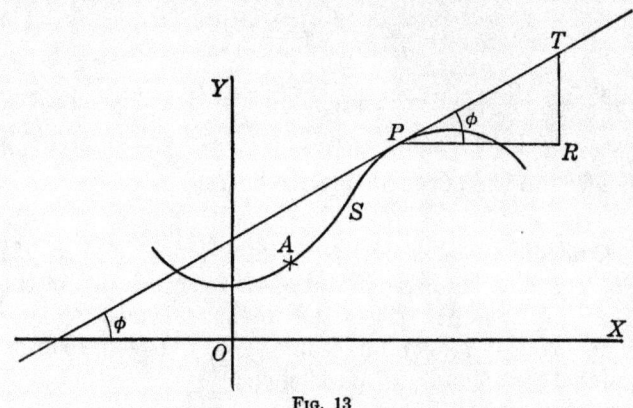

Fig. 13

tangent to the path at P making an angle ϕ with OX we can represent by PT the velocity of the particle at P, $\frac{ds}{dt}$, and by PR and RT the velocity components parallel to OX and OY respectively. We then have

$$PR = PT \cos \phi \quad \text{and} \quad RT = PT \sin \phi$$

or $\quad \dfrac{dx}{dt} = \dfrac{ds}{dt} \cos \phi \quad$ and $\quad \dfrac{dy}{dt} = \dfrac{ds}{dt} \sin \phi \qquad$ (16)

or $\quad v_x = v \cos \phi \quad$ and $\quad v_y = v \sin \phi$

which are the same relations as (15) since (Fig. 12), $\phi = 90° + \theta$, $\cos \phi = -\sin \theta$, $\sin \phi = \cos \theta$.

For convenience of reference the various formulas of Arts. 11 and 12 are here assembled.

$\dfrac{du}{dt}$ = rate of change of any variable u (with respect to time, t)

$\dfrac{ds}{dt} = v$ = velocity or speed.

$\dfrac{d^2s}{dt^2} = \dfrac{dv}{dt} = f$ = acceleration   (17)

$\dfrac{d\theta}{dt} = \omega$ = angular velocity.

$\dfrac{d^2\theta}{dt^2} = \dfrac{d\omega}{dt} = \alpha$ = angular acceleration.

$\left.\begin{array}{l} \dfrac{dx}{dt} = \dfrac{ds}{dt}\cos\phi \text{ or } v_x = v\cos\phi \\[2ex] \dfrac{dy}{dt} = \dfrac{ds}{dt}\sin\phi \text{ or } v_y = v\sin\phi \end{array}\right\}$ horizontal and vertical components of velocity.

$\left(\dfrac{ds}{dt}\right)^2 = \left(\dfrac{dx}{dt}\right)^2 + \left(\dfrac{dy}{dt}\right)^2 \text{ or } v^2 = v_x^2 + v_y^2.$

Example 1. A body moves along a straight line running East and West so that its distance from a fixed point 0 is given by the equation $s = t^3 - 3t^2 + 2$, s being distance and t, time. What is its speed at any instant? When $t = 3$? When will the body be stationary? When, how long and how far will it move east? West?

$v = \dfrac{ds}{dt} = 3t^2 - 6t$ and when $t = 0, v = 0$; when $t = 3, v = 9$.

When the body is stationary the speed is zero, therefore
$$3t^2 - 6t = 0, \quad t = 0, 2$$

Tabulating these results

t	0	1	2	3
$\dfrac{ds}{dt}$	0	—	0	+

Thus for the first two seconds $\frac{ds}{dt}$ is negative and since Δt is positive (time increases) s is decreasing and the body moves west. After two seconds $\frac{ds}{dt}$ is positive, s increases and the body moves east. The body starts at $s = 2$ (put $t = 0$ in $s = t^3 - 3t^2 + 2$). When $t = 1$, $s = 0$; therefore the body moves west a distance of two units.

Example 2. A spherical balloon is being filled with gas at the rate of 2 cubic feet per minute. How fast is the radius of the sphere increasing when the radius is 10 feet?

$$v = \tfrac{4}{3}\,\pi\,r^3$$

$$\frac{dv}{dt} = \tfrac{4}{3}\,\pi \,.\, 3\,r^2\,\frac{dr}{dt} = 4\,\pi\,r^2\,\frac{dr}{dt}$$

$$\frac{dv}{dt} = 2 \text{ cu. ft. min.}; \quad r = 10$$

$$\therefore 2 = 400\,\pi\,\frac{dr}{dt}; \quad \frac{dr}{dt} = \frac{1}{200\,\pi} \text{ ft. min.}$$

Example 3. The pressure times the volume of a gas is equal to 16. The volume is decreasing at the rate of 3 cu. cm. per second. At what rate is the pressure changing? When the volume is 10 cu. cm.?

$$pv = 16$$

$$v\frac{dp}{dt} + p\frac{dv}{dt} = 0$$

$$\frac{dv}{dt} = -3 \quad \therefore v\frac{dp}{dt} - 3\,p = 0 \quad \text{and}$$

$$\frac{dp}{dt} = \frac{3\,p}{v}$$

When $\qquad v = 10, \quad p = \dfrac{16}{v} = 1.6 \quad$ and

$$\frac{dp}{dt} = \frac{4.8}{10} = .48$$

Example 4. A metal cube, remaining cubical, expands under the influence of heat. How much faster does the volume increase than the edge?

$$v = x^3, \quad \frac{dv}{dt} = 3\,x^2\,\frac{dx}{dt} \quad \text{or} \quad \frac{\dfrac{dv}{dt}}{\dfrac{dx}{dt}} = 3\,x^2,$$

so that the volume is increasing $3\,x^2$ times as fast as the edge; in particular when $x = 2$, 12 times as fast; $x = 3$, 27 times as fast.

The interval of time being the same in both derivatives we may write *

$$\frac{\dfrac{dv}{dt}}{\dfrac{dx}{dt}} = \frac{dv}{dx} = 3\,x^2, \text{ the rate of change of the volume with respect}$$

to the edge of the cube. In general $\dfrac{dy}{dx}$ may mean the rate of change of y with respect to, or as compared with x. (18)

Example 5. A wheel of radius 2 feet makes 100 revolutions per minute. How fast is a point on the rim moving?

* In general $\dfrac{\Delta y}{\Delta x} = \dfrac{\dfrac{\Delta y}{\Delta z}}{\dfrac{\Delta x}{\Delta z}}, \quad \underset{\Delta x = 0}{\text{limit}} \dfrac{\Delta y}{\Delta x} = \underset{\Delta x = 0}{\text{limit}} \dfrac{\dfrac{\Delta y}{\Delta z}}{\dfrac{\Delta x}{\Delta z}} = \dfrac{\underset{}{\text{limit}} \dfrac{\Delta y}{\Delta z}}{\underset{}{\text{limit}} \dfrac{\Delta x}{\Delta z}}$

or $\dfrac{dy}{dx} = \dfrac{\dfrac{dy}{dz}}{\dfrac{dx}{dz}}.$ Thus $\dfrac{\dfrac{dv}{dt}}{\dfrac{dx}{dt}} = \dfrac{dv}{dx}.$ Similarly $\dfrac{dy}{dx} = \dfrac{dy}{dz} \cdot \dfrac{dz}{dx}.$

What are the x and y components of velocity when the wheel has moved through $\dfrac{3\pi}{4}$ radians?

Here $\dfrac{d\theta}{dt} = 100$ rev. min. $= 200\pi$ radians minute.

$$s = a\theta = 2\theta \quad \therefore \frac{ds}{dt} = 2\frac{d\theta}{dt} = 400\pi \text{ feet minute}$$

and is the same for all angles. Also at $\theta = \dfrac{3\pi}{4}$,

$$v_x = -\,400\pi \sin\frac{3\pi}{4} = -\,\frac{400\pi}{\sqrt{2}}$$

and $\ v_y = 400\pi \cos\dfrac{3\pi}{4} = \dfrac{400\pi}{\sqrt{2}}$.

Example 6. The power which turns the wheel of Example 5 is cut off, after which friction stops the wheel according to the law $\theta = 200\pi t - t^2$. Find the angular velocity and acceleration, the velocity and components of velocity of a point on the rim; at any instant and after 2 hours from the time the power is shut off. In what time will the wheel be brought to rest?

$$\omega = \frac{d\theta}{dt} = 200\pi - 2t \text{ rad. min.}$$

$$a = \frac{d^2\theta}{dt^2} = -\,2 \text{ rad. per min. per min.}$$

$$v = \frac{ds}{dt} = 2\frac{d\theta}{dt} = 400\pi - 4t \text{ ft. min.}$$

$$v_x = -\,v\sin\theta = -\,(400\pi - 4t)\sin\theta \text{ ft. min.}$$

$$v_y = v\cos\theta = (400\pi - 4t)\cos\theta \text{ ft. min.}$$

When $t = 2$ hr. these become (note, $t = 2$ hrs. $= 120$ min. gives $\theta = 24000\pi - 14400 = 60960$ rad. $= 3{,}493{,}008° = 9702 \times 2\pi + 288°$).

$$\omega = \frac{d\theta}{dt} = 200\,\pi - 240 = 388 \text{ rad. min.}$$

$$v = \frac{ds}{dt} = 400\,\pi - 480 = 776 \text{ ft. min.}$$

$$v_x = -776 \sin 288° = -776 \times (-.951) = 738 \text{ ft. min.}$$

$$v_y = 776 \cos 288° = 776 \times (\ \ .309) = 240 \text{ ft. min.}$$

The wheel will come to rest when

$$\frac{d\theta}{dt} = 200\,\pi - 2t = 0 \text{ or } t = 100\,\pi \text{ min.} = 5 \text{ hr. } 14 \text{ min.}$$

EXAMPLES

In the following examples in which s is distance in feet, t is time in seconds. Find (1) velocity, (2) acceleration, (3) starting point, (4) stationary points, (5) when and how long the particle moves N., S., S. W., etc., as the case may be, (6) how far it moves N., S., etc.

1. $s = 2 + 3t^2 - t^3$ Motion on a straight line running east and west.

2. $s = t^3 - 3t^2 + 2$ Straight line N. and S.

3. $s = 7 - 5t + 9t^2$ Straight line N. E. and S. W.

4. $s = 5 + 4t - t^2$ Straight line N. W. and S. E.

5. $s = t^3 - 3t^2$ Straight line E. and W.

6. $s = 3t^2 - t^3$ Straight line N. and S.

If a body falls near the surface of the earth its motion is represented by $s = 16t^2$ if it falls from a state of rest; by $s = 16t^2 - v_0 t$ if it had an initial upward velocity v_0, where s is in feet and t in seconds.

7. Find the velocity and acceleration of a body falling from rest, at the end of t seconds; 2 seconds, 5 seconds.

8. Find the velocity and acceleration of a falling body, with an initial velocity downward of 3 feet per second, at the end of t seconds; 3 seconds, 10 seconds.

9. A body at rest 1600 feet above the earth falls to the earth. With what velocity will it reach the earth's surface?

10. If the body of question 9 is thrown downward with a velocity of 20 feet per second, with what velocity will it reach the earth?

11. A body is thrown vertically upward with a velocity of 320 feet per second. How long and how high will it rise?

12. From what height must a body be dropped that it may strike the ground with a velocity of 200 ft. per second? In what time will it reach the ground?

13. With what velocity must a ball be thrown vertically upward to rise to a height of 320 feet?

14. What is the speed of a body falling from rest, at the end of 2 seconds? If this speed remained constant for one-tenth second how far would the body fall? How far does the body actually fall in the one-tenth second? What is the error, and the percentage error, in assuming the speed constant? Answer the same questions for one-hundredth of a second.

15. The ends of a trough 6 feet long are equal isosceles right triangles having the hypothenuse horizontal. A horse is drinking the water at the rate of a cubic foot per minute. How fast is the level of the water sinking when the water is 18 inches deep?

16. A trough is in the shape of a right prism, with its ends equilateral triangles. The length of the trough is 10 feet. It contains water, which leaks at the rate of 1 cubic foot per minute. Find in inches per second the rate at which the level is sinking when the depth of the water is 3 inches.

17. A vessel containing water is in the form of an inverted hollow cone with vertical angle 90°. If water flows in at the rate of 1 cubic foot per minute, at what rate is the level of the water rising when the depth is 2 feet?

18. In an hour glass, with vertical angle 90°, the sand is flowing from the upper cone to the lower at the rate of x cubic centimeters per second, where x is the height of the sand in the upper cone. How fast is the level of the sand sinking when the height of the sand is 1 centimeter? Does the level sink faster or slower as the sands run out?

19. A current C of electricity is changing according to the law $C = 20 + 21\,t - 14\,t^2$, where t is in seconds. The voltage V is such that $V = RC + L\dfrac{dC}{dt}$, where $R = 0.5$ and $L = 0.01$. Find V when $t = 2$. How fast is the voltage changing at any time? When $t = 2$? Is the voltage increasing or decreasing when $t = .73$ sec.? Why?

20. The coefficient of cubical expansion of a substance at temperature $\theta°$ is the rate of increase of volume per unit increase of temperature. The volume (c.c) of a gram of water at $\theta°$ C is given by $V = 1 +$ $a(\theta-4)^2$, when $a = 8.38 \times 10^{-6}$. Find $\dfrac{dV}{d\theta}$ and hence get the coefficient of cubical expansion of water at 0° C. and at 20° C.

21. The specific heat of a substance at temperature $\theta°$ is the rate of increase of Q per unit increase in θ, where Q is the number of heat units required to raise the temperature of 1 gm. from some standard temperature to θ. The total heat required to raise the temperature of 1 gm. of water from 0° to $\theta°$ C. is given by $Q = \theta + 2 \times 10^{-5} \theta^2 + 3 \times 10^{-7} \theta^3$. Find the specific heat of water at 80° C.

22. For a diamond the formula is (see Ex. 21)

$$Q = 0.0947\,\theta + 0.000497\,\theta^2 - 0.00000012\,\theta^3.$$

Find the specific heat of diamond at 80° C.

23. A wheel in t seconds rotates through $5t + 4t^3$ radians from some standard position. Find its angular velocity and acceleration after 5 seconds.

24. A fly wheel is making 100 revolutions per minute. When the power is cut off friction brings the wheel to rest according to the law $\theta = 200\,\pi t - 20\,\pi t^2$. Find the angular velocity and acceleration at any minute after the power is cut off; when $t = 2$ min. When will the wheel come to rest?

25. A wheel of radius 10 inches makes 100 revolutions per minute. Find the velocity components of a point in the rim one second after it crossed OX.

26. A particle moves in a curve given by the equations $x = 5\cos t$, $y = 5\sin t$, where t is the time. Prove that the path is a circle. (Give values to t, find x and y, and plot the points.) Find the x and y components of velocity when $t = 1$, $t = 2\pi$.

27. The path of a moving particle is given by $s = t^3 - 3t^2$. It is known that when $t = 3$ the particle is moving in a line making an angle of 45° with OX. Find the velocity and the velocity components when $t = 3$.

28. Which will come to rest sooner, a wheel revolving according to the law $\theta = t - t^2$ or one whose law is $\theta = t + 2\cos t$? How much sooner?

CHAPTER IV

APPLIED MEANINGS AND USES OF THE DERIVATIVE, CONTINUED. DIRECTION. MAXIMA AND MINIMA. DIFFERENTIALS, APPROXIMATIONS, ERRORS.

14. Direction of a Line. Slope. Let us consider a simple function $y = 2x - 4$, of which the graph is shown in Fig. 14 to be a straight line. Suppose we wish to know the direction of this straight line. To determine the direction

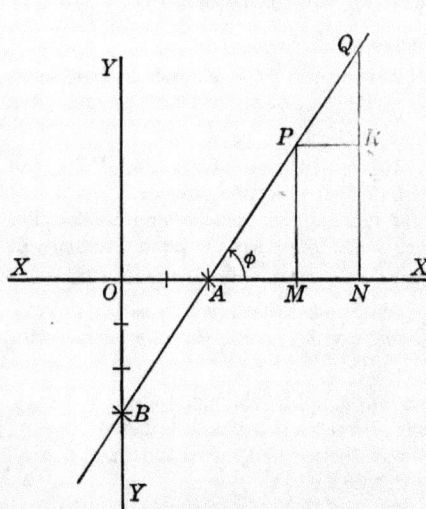

Fig. 14

we must have some reference line, just as in getting directions on the surface of the earth we use for reference the earth's axis and equatorial plane, and thus get directions northwest, southeast, etc. Let us choose the axis of x as our reference line and measure the direction of our line, $y = 2x - 4$, by means of the angle ϕ made by the line with OX, to the right of the line and above the axis. It is found convenient, in most cases, however, to use, instead of ϕ the tangent of ϕ. This is called the slope of the line. Thus

$$\text{Slope } BAP = \tan \phi \qquad (19)$$

46

Let $P\,(x,\,y)$ be any point and $Q\,(x + \Delta x,\, y + \Delta y)$ any other point on the line. Draw the coördinates of P and Q and draw PR parallel to OX. The angle RPQ is equal to ϕ. Therefore

$$\text{Slope } BAP = \tan \phi = \tan RPQ = \frac{RQ}{PR} = \frac{NQ - MP}{ON - OM} =$$
$$\frac{(y + \Delta y) - y}{(x + \Delta x) - x} = \frac{\Delta y}{\Delta x} \qquad (20)$$

so that the slope of a straight line may also be defined as the increment of the ordinate (Δy) divided by the increment of the abscissa (Δx).

Consider next the function $y = x^2$ of which the graph (Fig. 15) is a curved line, whose direction changes at every

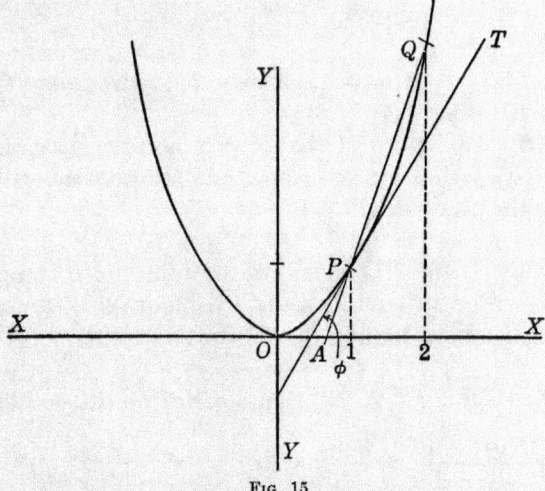

FIG. 15

point. What is meant by the direction of this line at the point P? As the point Q approaches nearer and nearer to P the secant PQ approaches nearer and nearer in position to the tangent line PT, and the angle XAP between the axis of x and the secant approaches nearer and nearer to the angle ϕ between the axis of x and the tangent line. By taking the point Q as near as we please to P we can make the position of the secant PQ, and the angle of the secant, approach as near as we please to the position and angle of the tangent line respectively.* Thus, to determine the direction of a curve we have

Slope of curve at P = slope of tangent line to curve at P = limit of slope of secant PQ as Q approaches P as a limit; or, in symbols,

$$\text{Slope of curve at any point} = \lim_{\Delta x = 0} \frac{\Delta y}{\Delta x} = \frac{dy}{dx} \qquad (21)$$

For the particular function sketched in Fig. 15

$$\text{Slope} = \frac{dy}{dx} = 2\,x.$$

At the point $x = 1$, $y = 1$, slope $= 2$; at the point $(5, 25)$ slope $= 10$, etc.

By virtue of (18) p. 41 the slope of a curve may also be considered as the rate of change of the ordinate with respect to the abscissa.

Example 1. Find the slope of the tangent to the curve $y = x^4 - x^2$ at any point and at the point $(2, 12)$; also the slope of the chord joining $(2, 12)$ to $(2.1, 15.04)$.

$$\frac{dy}{dx} = 4\,x^3 - 2\,x;\ \text{at}\ (2,\ 12),\ \text{slope} = 4(2)^3 - 2(2) = 28$$

$$\frac{\Delta y}{\Delta x} = \frac{15.04 - 12}{2.1 - 2} = \frac{3.04}{.1} = 30.4$$

Example 2. Find the slope of the curve $s = t^4 - 3$ at the point $t = 2$, $s = 13$.

* The student should show this by a carefully drawn figure.

$$\frac{ds}{dt} = 4\,t^3; \text{ at } (2, 13), \text{ slope } = 4(2)^3 = 32$$

It should be noted that if s represents distance and t time, the slope of the curve in this case is the speed of the moving body *in its path*.

EXAMPLES

Find the slope of each of the following curves at any point, and at the points named, and also the slope of the chord joining the named point to the point whose abscissa is greater by one-tenth.

1. $y = 2\,x^2 - 3$	(2, 5)	5. $xy = 10$	(2, 5)
2. $y = 4 - 3\,x^2$	(− 1, 3)	6. $x^2y = 20$	(2, 5)
3. $y = x^3 + 2\,x$	(2, 12)	7. $s = t^3 - 3\,t^2$	(3, 0)
4. $y = x^3 - x^2$	(3, 18)	8. $s = 3\,t^2 - t^3$	(3, 0)

Has the particle moving in the path given below (s = distance, t = time) a greater speed when $t = 1$ or when $t = 2$? Verify your result by sketching the graph.

9. $s = 3\,t^2 - 4\,t$	10. $s = t^2 - \frac{t}{4}$	11. $st = 12$
12. $s = 1 - t^3$	13. $s = t^3 - 4$	14. $s = t - t^2$

Find the slope of each of the following curves at the points named.

15. $y = \sin 2\,x$, $\qquad x = \frac{\pi}{2}, \frac{\pi}{4}, \frac{\pi}{6}$. $\qquad$ 18. $y = \tan \frac{x}{2}$, $\qquad x = 0, \frac{\pi}{2}$.

16. $y = \cos 2\,x$, $\qquad x = \frac{\pi}{2}, \frac{\pi}{4}, \frac{\pi}{6}$. $\qquad$ 19. $y = \sec 3\,x$, $\qquad x = 0, \frac{\pi}{2}$.

17. $y = \tan x$, $\qquad x = 0, \frac{\pi}{4}$. $\qquad$ 20. $y = \csc \frac{x}{3}$, $\qquad x = 0, \frac{\pi}{2}$.

15. Maxima and Minima. Consider the function $y = -\frac{x^3}{2} + \frac{3\,x^2}{2}$ (Fig. 16). For different values of x, the function, y, takes different values, sometimes increasing and sometimes decreasing. Suppose we ask the question: What is the greatest value of the function for the range of values of x from − 1 to 3? By (18) p. 41, $\frac{dy}{dx}$ is the rate of change of y with respect to x, and by pp. 34–35 (assuming that x continuously increases) the function y increases when $\dfrac{dy}{dx}$ is positive, decreases when $\dfrac{dy}{dx}$ is negative. The function ceases to increase and begins to decrease when $\dfrac{dy}{dx}$ changes from being positive to being negative; that is, when $\dfrac{dy}{dx}$ is equal to zero. Let us inquire at the same time, what is

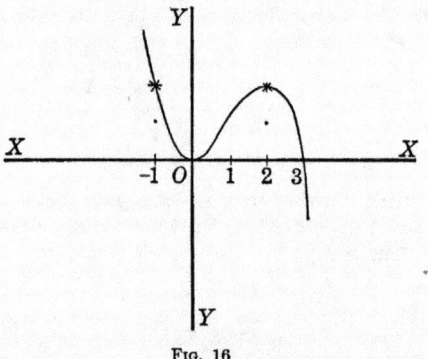

Fig. 16

the least value of the function between $x = -1$ and $x = 3$? Obviously y decreases when $\dfrac{dy}{dx}$ is negative; increases when $\dfrac{dy}{dx}$ is positive; ceases to decrease and begins to increase when $\dfrac{dy}{dx}$ changes from being negative to being positive, or when $\dfrac{dy}{dx} = 0.^*$ In either case, for the function to reach a maximum value or a minimum value, must $\dfrac{dy}{dx} = 0.$ This, as we say, is a *necessary* condition for a maximum or

* Or when $\dfrac{dy}{dx} = \infty$. This case will not be considered.

minimum value of the function. But it is not a *sufficient* condition; that is, it is not enough, because even if $\dfrac{dy}{dx} = 0$ we may have either a maximum or a minimum value of the function y or may have neither the one nor the other. Let us now proceed with the example. Placing

$$\frac{dy}{dx} = -\frac{3\,x^2}{2} + 3\,x \text{ equal to zero,}$$

$$-\frac{3\,x^2}{2} + 3\,x = 3\,x\left(1 - \frac{x}{2}\right) = 0$$

we obtain $\qquad x = 0 \quad \text{and} \quad x = 2$

and we may tabulate our information as follows:

x	-1	0	1	2	3
$\dfrac{dy}{dx}$	$-$	0	$+$	0	$-$
y	Decreasing	0 Min. Value	Increasing	2 Max. Value	Decreasing

That is, beginning with $x = -1$ and ending with $x = 3$, for values of x less than 0, y decreases; for values of x between 0 and 2, y increases; for values of x greater than 2, y decreases. Therefore the least value of the function (a minimum) is $y = 0$, when $x = 0$, and the greatest value (a maximum) is $y = 2$, when $x = 2$.

It should be noted that a $\begin{cases} \text{maximum} \\ \text{minimum} \end{cases}$ value is not necessarily the $\begin{cases} \text{greatest} \\ \text{least} \end{cases}$ value a function may have. For instance in the above function, $x = -3$ gives $y = 27$ which is greater than the maximum value $y = 2$, and $x = 4$ gives $y = -8$ which is less than the minimum value $y = 0$. A $\begin{cases} \text{maximum} \\ \text{minimum} \end{cases}$ value is one which is $\begin{cases} \text{greater} \\ \text{less} \end{cases}$ than the values immediately before and immediately after. It will be seen (Fig. 17) that at a maximum or minimum point

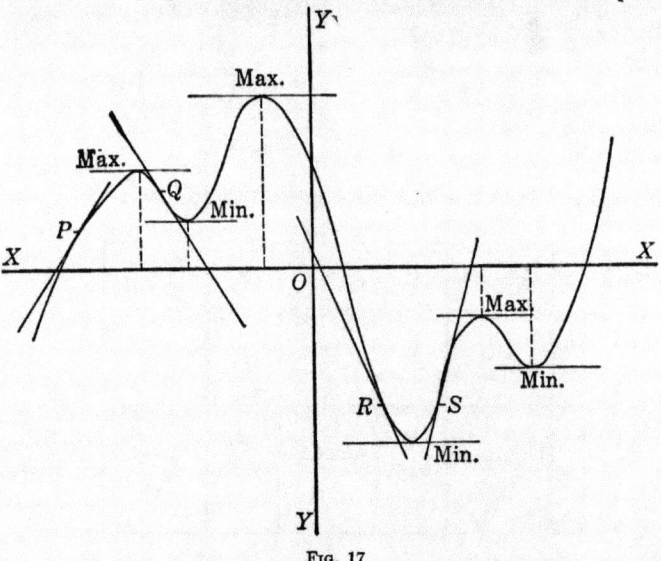

Fig. 17

the tangent to the curve is parallel to the axis of x, so that the slope is zero; that is, $\frac{dy}{dx} = 0$. Just before a maximum (point P) the tangent makes an acute angle with OX and $\frac{dy}{dx} = +$; just after a maximum (point Q) the tangent makes an obtuse angle with OX and $\frac{dy}{dx} = -$. Just before a minimum (R) the angle is obtuse and $\frac{dy}{dx} = -$; just after a minimum (S) the angle is acute and $\dfrac{dy}{dx} = +$. These results are identical with those obtained from the point of view of increasing and decreasing values of the function.

We may summarize the results as follows:

For a $\begin{cases} \text{maximum} \\ \text{minimum} \end{cases}$ value of y, a function of x, $\dfrac{dy}{dx} = 0$ and changes from $\begin{cases} + \\ - \end{cases}$ to $\begin{cases} - \\ + \end{cases}$ as x increases.* . . . . (22)

Example 1. Find maximum and minimum values of the function $2\,x^3 - 3\,x^2 - 12\,x + 6$.

Put $u = 2\,x^3 - 3\,x^2 - 12\,x + 6$

$$\dfrac{du}{dx} = 6\,x^2 - 6\,x - 12 = 6\,(x+1)\,(x-2) = 0$$

$$x = -1,\, 2$$

x	-2	-1	0	2	3
$\dfrac{du}{dx}$	$+$	0	$-$	0	$+$
u	incr.	13 Max.	decr.	-14 Min.	incr.

That is, the maximum value of the function is 13, when $x = -1$; the minimum value is -14, when $x = 2$.

Example 2. Find maximum and minimum distances from the origin of a point which moves in a straight line according to the law $s = t^3 - 12\,t^2 + 45\,t + 3$ (feet; seconds).

* When $\dfrac{dy}{dx} = 0$ and does not change sign, we can only say that y has neither a maximum nor a minimum value.

$$\frac{ds}{dt} = 3\,t^2 - 24\,t + 45 = 3\,(t-3)\,(t-5) = 0$$

whence $t = 3$ or 5.

t	0	3	4	5	6
$\dfrac{ds}{dt}$	$+$	0	$-$	0	$+$
s	incr.	57	decr.	53	incr.

that is, the maximum distance is 57 feet at the end of 3
seconds; the minimum, 53 feet at the end of 5 seconds.
This means that the point, starting at $s = 3$ feet, moves
away from the origin until, at the end of 3 seconds, it is
57 feet away. Then it moves toward the origin until, at
the end of 5 seconds, it is 53 feet away. After that it moves
away from the origin indefinitely. The maximum s is not
the greatest distance, for $t = 7$, for example, gives $s = 73$.
The minimum s is not the least distance, for $t = 0$ gives
$s = 3$.

Example 3. A sheet of zinc 6 feet long and 3 feet wide
is to be made into a rectangular tank by cutting out square
corners and folding up the edges. Find the dimensions of
the tank when its capacity is a maximum. See Fig.

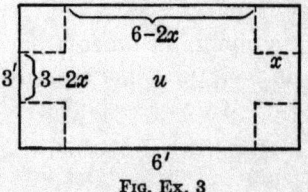

Let x be a side of the square
cut out and u the volume of
the tank. Then $u = x\,(6 - 2\,x)$
$(3 - 2\,x)$

or $u = 2\,(2\,x^3 - 9\,x^2 + 9\,x)$

FIG. EX. 3

$$\frac{du}{dx} = 2 (6 x^2 - 18 x + 9) = 0$$

$$x = \frac{3 \pm \sqrt{3}}{2} = 2.37 \text{ or } .64 \text{ feet}$$

The first value is obviously impossible as it is more than half the shorter side of the sheet of zinc. Therefore if there be a maximum value it will occur when $x = .64$ feet. There must be a maximum volume since when $x = 0$ the volume is zero and increases with x. When $x = 1.5$ feet the volume is zero and increases as x decreases. The volume increasing from both extremes must reach a maximum value In most practical problems it is obvious that there must be a maximum or minimum, as the case may be, and no further test need be applied. In the above problem, therefore, the dimensions for a maximum are length 4.72 feet, width 1.72 feet, volume 5.2 cubic feet.

Example 4. Find maximum and minimum velocities of a particle moving in a straight line according to the law $s = 3 t^3 - 2 t^2$.

$$\frac{ds}{dt} = 9 t^2 - 4 t = v.$$

$$\frac{dv}{dt} = 18 t - 4 = 0, t = \tfrac{2}{9}$$

t	0	$\tfrac{2}{9}$	1
$\frac{dv}{dt}$	—	0	+
v	decr.	Min. $-\tfrac{4}{9}$	incr.

Thus $t = \frac{2}{9}$ gives $v = -\frac{4}{9}$, a mathematical minimum. This is not the least speed, however, since obviously, v is numerically least when $t = 0$. There is no maximum.

Example 5. Find the dimensions of the greatest right circular cylinder inscriptible in a right circular cone of radius r and height h, the planes of the bases of the cone and cylinder being the same. See Fig.

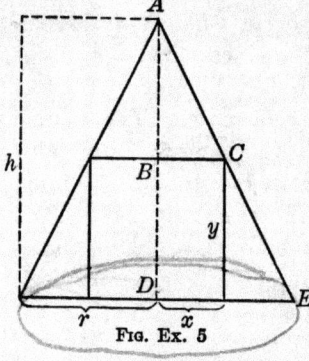

FIG. EX. 5

Let v be the volume of the cylinder. Then $v = \pi x^2 y$ which is a function of two variables x and y. These are not, however, independent, since the triangles ABC and ADE being similar,

$$\frac{AB}{BC} = \frac{AD}{DE}; \text{ or, } \frac{h-y}{x} = \frac{h}{r} \text{ and } y = \frac{h}{r}(r-x).$$

Therefore, $v = \dfrac{\pi h}{r} x^2 (r-x) = \dfrac{\pi h}{r}(rx^2 - x^3)$

$$\frac{dv}{dx} = \frac{\pi h}{r}(2rx - 3x^2) = 0$$

$$x = 0 \text{ or } \frac{2r}{3}$$

The first value is obviously impossible, and the second obviously gives a maximum cylinder whose dimensions are:

$$x = \frac{2r}{3}, \; y = \frac{h}{3}, \; v = 4\frac{\pi r^2 h}{27} = \left(\frac{4}{9} \text{ volume of cone.}\right)$$

Example 6. A tin tomato can is to be made in the form of a right circular cylinder. What will be its dimensions when the amount of tin used is least? See Fig.

Let u be the total surface of the can.

Then

$$u = 2 \pi r^2 + 2 \pi rh$$

a function of two variables. No data are given by which we can connect r and h, but common sense tells us that the can must be of a definite capacity, so that we may write for the volume of the can

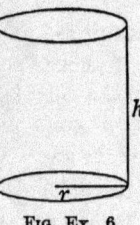

FIG. EX. 6

$$v = \pi r^2 h = \text{constant} = a.$$

From this the value of h might be expressed in terms of r and substituted in u, as in Example 5; or we may proceed as follows:

$$\frac{du}{dr} = 2 \pi \left\{ 2r + h + r \frac{dh}{dr} \right\} = 0 \text{ and}$$

$$\frac{dv}{dr} = \pi \left\{ 2rh + r^2 \frac{dh}{dr} \right\} = \frac{da}{dr} = 0$$

From the latter equation $\frac{dh}{dr} = -\frac{2h}{r}$

which being substituted in the former gives

$$2r + h - 2h = 0$$

or $h = 2r$.

That is, for a cylindrical can of any definite capacity, the amount of material will be least when the height is equal to the diameter of the base.

EXAMPLES

Find maximum and minimum values of the following functions. Sketch the graphs.

1. $x^3 + 2 x^2$ 2. $x^2 - 3 x^3$ 3. $x + x^2 + x^3$
4. $2 x - x^2 + x^3$ 5. $3 - 2 x^3$ 6. $4 x - 3 x^3$
7. $\sin 3 x$ 8. $\sin x + \cos x$ 9. $\sin x - \cos x$

In the following examples the path of a particle moving in a straight line is given, s being distance in feet, t time in seconds. In each case find the greatest or least speed of the particle during the first second.

10. $s = 3 t^3 - 2 t^2 + 3 t$ 12. $s = 16 t - 2 t^2 + t^3$
11. $s = t + t^2 - t^3$

13. Find the dimensions of the rectangle of greatest area with a perimeter of 10 feet. Draw a graph showing how the area changes with the length of the rectangle.

14. The surface of a hollow cylinder without top is 100 square inches. Find the maximum volume.

15. The volume of a solid cylinder is V cubic inches. Find its dimensions if the total surface is a minimum.

16. A running track is in the form of a rectangle $ABCD$ with semicircles on AB, CD. If the exterior perimeter is a quarter of a mile, find the maximum area inclosed.

17. A window is in the form of a rectangle surmounted by a semicircle. If the perimeter is 30 feet, find the dimensions so that the greatest possible amount of light may be admitted.

18. A shot is projected *in vacuo* with velocity u feet per second in a direction making an angle a with the horizontal. Its height above the ground at the end of t seconds is $tu \sin a - \frac{1}{2} g t^2$. Find the greatest height and the time of reaching it.

19. Find the circular cylinder of largest volume which can be cut from a sphere of radius 6 inches, the plane ends being perpendicular to the axis.

20. Given 200 square feet of canvas, find the greatest conical tent that can be made out of it.

21. Given a circular sheet of paper, find the angle of the sector which must be cut out so that the remainder may be folded to give a conical vessel of maximum volume.

22. Find the least area of canvas that can be used to construct a conical tent whose capacity is 800 cubic feet.

23. Find the volume of the greatest right cone that can be described by the revolution, about a side, of a right triangle of hypotenuse 2 feet.

24. The annual cost of giving a certain amount of electric light to a town, the voltage being V and the candle-power of each lamp C, is found to be

$$A = a + \frac{b}{V} \text{ for electric energy,}$$

and

$$B = \frac{m}{C} + \frac{nV^{2/3}}{C^{5/3}} \text{ for lamp renewals.}$$

The following figures are known when C is 10.

V	100	200
A	1500	1200
B	300	500

Find a, b, m, n. If $C = 20$, what value of V will give the minimum total cost?

25. In measuring electric current by a tangent galvanometer the percentage error due to a given small error in the reading is proportional to $\tan x + \cot x$. Find the value of x for which this is a minimum. (Put $\tan x = t$.)

26. An electric current flows around a coil of radius a. A small magnet is placed with its axis on the line perpendicular to the plane of the coil through its center. If x is the distance of the magnet from the plane of the coil, the force exerted on it by the current is proportional to

$$\frac{x}{(x^2 + a^2)^{5/2}}.$$

Find x so that the force may be a minimum. (Put $x^2 + a^2 = y$.)

27. A ship, P, is 90 miles south of another ship, Q. P is sailing north 15 miles per hour, and Q east 12 miles per hour. How long will the ships approach each other? When will they be nearest?

28. The signals between two boats can be read at a distance of one mile. A boat, B, sailing due south at 6 knots lies 5 miles due west of a boat, C, which is sailing west at 7 knots. Do the boats come within signaling distance? When? Why?

29. A brace in the form of the letter Y is to be 16 feet in total height and 12 feet in width across the top. Find the length of the stem and the length of an arm if the length of the stem plus that of the two equal arms is a minimum.

30. A rectangular garden, to contain 4000 square feet, is to be laid out on the boundary between two lots and to be fenced in. If the cost of the fence along the boundary is to be shared equally by the two abuttors, what must be the dimensions of the garden that the total cost of fencing to the owner of the garden may be least?

31. A window in the form of a rectangle surmounted by an equilateral triangle is 35 feet in perimeter. Find its dimensions to admit the maximum amount of light.

32. A wire 10 feet long is cut into two pieces. One piece is made into a circle; the other into a square. Find the diameter of the circle and the side of the square when the total area is a minimum.

33. A tank in the form of a right circular cylinder, open at the top, is to hold 5000 cubic feet. The bottom costs $2 per square foot; the sides $1.50 per square foot. Find the dimensions for the minimum cost.

34. A billboard 6 feet high stands on posts 10 feet high. How far from the edge of the road must the board be placed that it may subtend the greatest angle at the eye of a person 5 feet 6 inches tall? (Suggestion: Find maximum value of tan θ, when θ is the subtending angle.)

16. Differentials. Approximate Increments. Errors.

If y be a function of x it is obvious that

$$\Delta y = \frac{\Delta y}{\Delta x} \cdot \Delta x \qquad (23)$$

even when Δy and Δx are small numbers. If we allow Δx to become very small (to approach zero) then $\frac{\Delta y}{\Delta x}$ approaches $\frac{dy}{dx}$ as a limit. To indicate that Δx and Δy are thus becoming very small we shall use the notation dx and dy, which are called differentials, and shall write

$$dy = \frac{dy}{dx}\, dx \qquad (24)$$

That is

the differential of y equals the derivative of y with respect to x times the differential of x.

We have not hitherto used $\frac{dy}{dx}$ as a fraction but (24) above shows that if we use it as a fraction and cancel the dx the result is the differential of y as defined above.

It is obvious that the increment of y (Δy) and the differential of y (dy) are not equal, but it is also obvious that the smaller we make the increment of x (Δx) the nearer will Δy and dy come to being equal. For example let $y = x^2$, then

$$\Delta y = (x + \Delta x)^2 - x^2 = 2 x \Delta x + \overline{\Delta x}^2$$

But
$$2 x = \frac{dx^2}{dx} = \frac{dy}{dx}, \text{ so that}$$

$$\Delta y = \frac{dy}{dx} \Delta x + \overline{\Delta x}^2$$

If now the value of Δx be taken very small, the value of $\overline{\Delta x}^2$ becomes very small even as compared with Δx; * so small that we neglect it for practical purposes and write, when Δx or dx becomes smaller than any number we please to assign, the differential of y, in general,

$$dy = \frac{dy}{dx} dx$$

and for the function $y = x^2$

$$dy = 2 x dx.$$

We thus have a new idea (the differential) and a new notation (the differential notation) which in many cases are more useful and convenient than the derivative and the derivative notation. The formulas, (6) p. 26 may now be written

* Let $\Delta x = .00001$, for example. Then $\overline{\Delta x}^2 = .0000000001$.

$$d(u + v + w + ...) = du + dv + dw + ...$$

$$d(uv) = v\,du + u\,dv$$

$$du^n = nu^{n-1}\,du$$

$$d(c\,u) = c\,du$$

$$d\sin u = \cos u\,du \tag{25}$$

$$d\tan u = \sec^2 u\,du$$

$$d\sec u = \sec u \tan u\,du$$

$$d\cos u = -\sin u\,du$$

$$d\cot u = -\csc^2 u\,du$$

$$d\csc u = -\csc u \cot u\,du$$

It should be noted that x being the independent variable we can give its increment, Δx, any value we please large or small, and we use dx merely for convenience to mean the value of Δx when it approaches zero. We can not say that Δy and dy are the same. It is true, however, that in the case of continuous functions, the kind with which we deal, Δy approaches zero as Δx does, and in dealing with differentials we are using quantities that approach zero as a limit, infinitesimal quantities, as they are called.

Consider the equation $y = x^2$ which may be used to express the area of a square as a function of the side. Let $x = 3$, $y = 9$. Increase x by .1; then $y + \Delta y = (3.1)^2 = 9.61$ and $\Delta y = 9.61 - 9 = .61$. If, however, we increase Δx by .01, we find $\Delta y = .0601$. Let us write the differential of y,

$$dy = 2\,x\,dx$$

If we put $dx = .1$, x being 3, we have $dy = .6$; $dx = .01$, $dy = .06$. Thus

x	Δx	Δy	dy	$\Delta y - dy$
3	.1	.61	.6	.01
3	.01	.0601	.06	.0001

and the smaller Δx is taken the nearer will the values of Δy and dy be to each other. In many problems the error arising from the use of dy instead of Δy is so small as to be negligible to the degree of accuracy with which we are working. For example, in the above function the error when

$$dx = .1 \text{ is } \Delta y - dy = .01 \text{ or } \frac{.01}{.61} = 1.6\%$$

When $dx = .01$ the error is $\Delta y - dy = .0001$ or $\frac{.0001}{.0601} = .17\%$.

We may illustrate the difference between Δy and dy in the above function graphically. In Fig. 18, $ABGK$ is the original value of y; $ACEH$ the new value; Δy is $BCDG + GDEF + GFHK$. But dy is $BCDG + GFHK$, and $\Delta y - dy$ is $DEFG$.

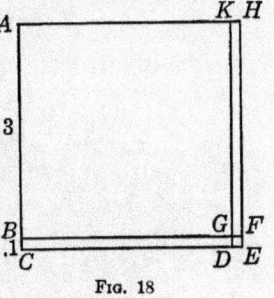

Fig. 18

Example 1. Find the differential of the function $3 x^4 + \sqrt{x^2 - 3}$.

Let $u = 3 x^4 + \sqrt{x^2 - 3} = 3 x^4 + (x^2 - 3)^{1/2}$

Then

$$du = \frac{du}{dx} \cdot dx = \frac{d\{3 x^4 + (x^2 - 3)^{1/2}\}}{dx} \cdot dx$$

$$= \left\{ 12\,x^3 + \tfrac{1}{2}\,(x^2 - 3)^{-1/2} \cdot 2\,x \right\} dx$$

$$= \left(12\,x^3 + \frac{x}{\sqrt{x^2 - 3}} \right) dx.$$

Example 2. Find the differential of $\sin^2 3\,x$.

$$d\,(\sin^2 3\,x) = 2 \sin 3\,x\, d\,(\sin 3\,x)$$

$$= 2 \sin 3\,x \cdot \cos 3\,x\, d\,(3\,x)$$

$$= 2 \sin 3\,x \cos 3\,x \cdot 3\,dx$$

$$= 6 \sin 3\,x \cos 3\,x dx = 3 \sin 6\,x dx.$$

Example 3. A circle of radius 3 feet has its radius increased by .02 feet. Find approximately the increase in area of the circle.

Here $A = \pi\,r^2$,

$$dA = 2\,\pi\,r dr \text{ and when } x = 3, \ dr = .02,$$

$$dA = (6\,\pi)\,(.02) = .12\,\pi \text{ square feet.}$$

Example 4. In the circle of example 3 what is the percentage error made in using dA as the increase in area instead of ΔA the actual increase?

$$\Delta A = \pi\,(3.02)^2 - \pi\,(3)^2 = \pi\,(\overline{3.02}^2 - 3^2) = .1204\,\pi$$
$$\text{square feet.}$$

$$\Delta A - dA = .1204\,\pi - .12\,\pi = .0004\,\pi \text{ square feet.}$$

$$\frac{\Delta A - dA}{\Delta A} = \frac{.0004\,\pi}{.1204\,\pi} = .003 = .3\%$$

Example 5. Compute, approximately, the volume of a cube whose edge is 3.004 feet.

Let $v = x^3$, then $dv = 3\,x^2 dx$.

When $x = 3$ and $dx = .004$, then

$$dv = (3)\,(9)\,(.004) = .108 \text{ cubic feet.}$$

So that $v' = v + dv = 27.108$ cubic feet.

Example 6. Find, approximately, the value of $\sin x + 2 \cos x$ when $x = 46°$.

Let $u = \sin x + 2 \cos x$ and, when $x = 45°$,

$$u_1 = \sin 45° + 2 \cos 45° = \frac{3\sqrt{2}}{2} = 2.121.$$

Also $du = (\cos x - 2 \sin x)\,dx$, and when $x = 45°$,

$$dx = 1° = .017 \text{ radians,}$$

$$du = (\cos 45° - 2 \sin 45°)\,(.017) = - \frac{.017\sqrt{2}}{2}.$$

Therefore the value of $\sin x + 2 \cos x$ when $x = 46°$ is

$$u_2 = \frac{3\sqrt{2}}{2} - \frac{.017\sqrt{2}}{2} = \frac{2.983\sqrt{2}}{2} = 2.109.$$

EXAMPLES

Find the differential of each of the following functions.

1. $x^3 - 3\,x^2$ 2. $x + x^2 - x^4$ 3. $x^3 - x^{3/2} + 4$

4. $\sqrt{x^3 + x}$ 5. $\sqrt[3]{x^2 - 2}$ 6. $(x^2 + 1)^{2/3}$

7. $\sin^2 3\,x$ 8. $\tan 5\,x$ 9. $\sec^2 \frac{x}{2}$

10. $x + \sin x$ 11. $\cot (3\,x + \frac{\pi}{4})$ 12. $\cos (2\,x + \frac{\pi}{6})$

Find dy from each of the following functions:

13. $xy = 20$ 14. $x^2 y = 10$ 15. $xy^2 = 5$

16. $x^2 + xy + y^2 = 2$ 17. $x^2 + y^2 = 16$ 18. $2\,x^2 - 3\,y^2 = 6$

19. Find approximately the volume of a thin spherical shell, internal radius 1 foot, thickness .2 inch. (Show that your result is less than 2 per cent in error.)

20. A stone is thrown at an angle of 40° to the horizontal with a velocity of 80 feet per second. Find the range $\left(R = \dfrac{V^2 \sin 2\,\theta}{g} \right)$ and the approximate increase in the range if the velocity is changed to 82 feet per second, the angle of projection remaining unchanged.

21. A cylindrical well is said to be 25 feet deep and 6 feet in diameter. Find the error in the calculated volume if there is an error of (1) 1 inch in the diameter, (2) 3 inches in the depth.

22. The radius of the base of a cone is r and its vertical angle $2\,a$. Find the approximate increase in volume due to a small increase $\Delta\,r$ in the radius, the vertical angle remaining constant. Hence show that the volume of a conical shell, internal radius r, thickness t, t being small, is approximately $\pi\,rlt$ where l is the slant height.

23. The radius of a sphere is found by measurement to be 18.5 inches, with a possible error of .1 inch. Find the consequent errors possible in (1) the surface area, (2) the volume, calculated from this measurement.

24. The side of a square is measured and found to be 8 inches. If an error of 0.01 inch is made in measuring the side, find approximately the error in the calculated area.

25. If the edge of a cube is measured and found to be 8 inches, and if an error of $1/_{20}$ inch has been made, what is the approximate error in the calculated volume?

26. Find approximately the value of $7\,x^2 - 3\,x + \dfrac{4}{x}$ when $x = 2.01$.

27. Find approximately the value of
$$2\,x^3 - 9\,x^2 + 12\,x - 3 \text{ when } x = (1)\ 2.005,\ (2)\ 1.995$$
In each case find also the exact value.

28. The annual cost for electric energy in a certain plant is $C = a + \dfrac{b}{V}$ where V is the voltage and a and b are constants. It is observed that $a = 900$ and $b = 60000$. If, when $V = 100,000$ an error of 1000 is made in V what is the approximate error in C?

29. A right triangle by measurement has sides $a = 300$ $b = 400$ and $c = 500$ feet long respectively. Find approximately

(a) The error in the area of the triangle if there is an error of 5 inches in measuring a, b being exact.

(b) The error in the area for an error of 5 inches in b, a being exact.

(c) The error in the area when a and b are each too small by 5 inches.

30. A right triangle has legs of $a = 300$ and $b = 400$ feet in length by measurement. Find the approximate error in the length of the hypothenuse when

(a) The legs have each a positive error of one per cent.

(b) Each a negative error of one per cent.

(c) Leg a a positive and b a negative error of one per cent.

31. The resistance of a circuit was found by using the formula $C = E/R$, where C = current and E = electromotive force. If there is an error of $1/10$ ampere in reading C and $1/20$ volt in reading E, what is the error in R if readings are $C = 20$ amperes and $E = 120$ volts? What is the maximum percentage error?

32. If the formula $\sin (x + y) = \sin x \cos y + \cos x \sin y$ were used to calculate $\sin (x + y)$, what approximate error would result if an error of 0.1 were made in measuring both x and y, the measurements of the two acute angles giving $\sin x = 3/5$ and $\sin y = 5/13$?

33. The acceleration of a particle down an inclined plane is given by $f = g \sin \alpha$. If α, which is measured as 30°, may be in error .01°, what is the maximum error in f? Take $g = 32$ ft. sec².

34. The period of a pendulum is $P = 2 \pi \sqrt{\dfrac{L}{g}}$. What is the greatest error in the period if there is an error of $\pm \frac{1}{10}$ foot in measuring a 10-foot suspension? What is the percentage error? (g equals 32 ft. sec²).

35. The length L and the period P are connected by the equation $4 \pi^2 L = P^2 g$. If L is calculated assuming $P = 1$, and $g = 32$ ft. sec², what is the approximate error in L if the true value is $P = 1.02$? What is the percentage error?

36. A particle is moving on the ellipse $x^2 + 4 y^2 = 20$ and is at the point (2, 2). What is the approximate change in y due to a change of 0.1 in x?

37. The force of attraction between two equal masses is $F = \dfrac{m^2}{s^2}$, where m is the value of each mass and s the distance between them. If m is measured as 10 grammes and s as 5 centimeters, find the error in F when

(a) s is exact and $dm = \pm .01$ gr.

(b) m is exact and $ds = \pm .01$ cm.

(c) $dm = .01$ gr. and $ds = .01$ cm.

(d) $dm = .01$ gr. and $ds = - .01$ cm.

CHAPTER V

INTEGRATION. INDEFINITE INTEGRALS. METHODS OF INTEGRATION

17. Integration. The Indefinite Integral. In all our work hitherto we have had given, y, a function of an independent variable x, and have asked what is the derivative (or differential) of y with respect to x. We have also learned how to obtain the derivative. Let us now assume that the derivative is given and ask what is the function of which it is the derivative? For example if we have the function $y = x^2$ we know that $\frac{dy}{dx} = 2x$ or that $dy = 2x\,dx$.* It is obvious that we can write

The function of which $2x\,dx$ is the differential is x^2 simply by using inversely, backwards, so to speak, our knowledge of the result of differentiation.

Our next thought would be to devise some simple way of making our statement in symbols instead of words, and the notation used is as follows:

$$\int 2x\,dx = x^2.$$

Thus $\int 2x\,dx = x^2$ states in symbols that the function of which $2x\,dx$ is the differential is x^2; more exactly that it *may be* x^2. It might equally well be $x^2 + 2$ or $x^2 - 5$ or $x^2 +$ any constant whatever, since the derivative of $x^2 + c$ is $2x$. The complete statement is, therefore,

$$\int 2x\,dx = x^2 + c \qquad (26)$$

---

* In this chapter we shall use differentials and the differential notation as being more convenient.

68

Having introduced this new symbol we give a name to it, namely integral, and the expression (26) is read:

The integral of $2x\,dx$ (or the integral of $2x$ with respect to x) is $x^2 + c$. The process of finding the integral is called integration.

18. Methods of Integration. Recognition, Transformation, Substitution. There are various methods of integration, of which we shall use here only three,* namely:

I. Recognition. II. Transformation. III. Substitution.

The first method we have already used in the above example, $\int 2x\,dx = x^2 + c$. It consists simply in recognizing the expression to be integrated, the integrand as it is called, as the differential of a certain function. The recognition, however, can be generalized and the process simplified by introducing certain theorems or formulas of integration. Thus we know that

$$d\,(u + v + w + \ldots) = du + dv + dw + \ldots$$

But the function of which $d\,(u + v + w + \ldots)$ is the differential is obviously $u + v + w + \ldots$, and this is the function of which du is the differential plus the function of which dv is the differential, etc. In symbols †

$$\int (du + dv + dw + \ldots) = \int du + \int dv + \int dw + \ldots \quad (27)$$

Stated briefly in words, " the integral of a sum equals the sum of the integrals of the terms."

In a similar way it may be shown that, a being a constant,

$$\int a\,du = a \int du \quad (28)$$

---

* A fourth method, integration by parts, is explained in Art. 34 following.

† The constant of integration will, in general, be omitted. It is always understood.

In words, " a constant factor can be put outside of the integral sign."

Again we know that

$$dx^n = n\, x^{n-1}\, dx$$

Therefore,

$$\int n x^{n-1}\, dx = \int dx^n$$

or by (28)

$$n \int x^{n-1}\, dx = x^n$$

$$\int x^{n-1}\, dx = \frac{x^n}{n}$$

which can be put in a rather better form by writing $n - 1 = m$ and $n = m + 1$, thus

$$\int x^m\, dx = \frac{x^{m+1}\,*}{m + 1} \tag{29}$$

The above formulas are grouped here for convenience of reference, together with four others immediately recognizable.

i. $\displaystyle\int (du + dv + dw + \ldots) = \int du + \int dv + \int dw + \ldots$

ii. $\displaystyle\int a\, du = a \int du$ $\tag{30}$

iii. $\displaystyle\int u^m\, du = \frac{u^{m+1}\,\dagger}{m + 1}$

iv. $\displaystyle\int \sin u\, du = -\cos u$ v. $\displaystyle\int \cos u\, du = \sin u$

vi. $\displaystyle\int \sec^2 u\, du = \tan u$ vii. $\displaystyle\int \csc^2 u\, du = -\cot u$

The second method of integration is called *transformation*. It consists merely in changing the form of the integrand,

---

* This formula fails when $m = -1$.
† This formula fails when $m = -1$.

the expression to be integrated, so that we can make use of the first method; can recognize the integrand as coming under one or more of the forms of (30). No rules can be given for the transformation to be employed, as we can use any algebraic or trigonometric transformation that may seem available. One or two examples will illustrate the process. It should be borne in mind that the object of the transformation is to obtain one or more forms that can be recognized.

Example 1. $\int \dfrac{x^2 + 3x + 2}{\sqrt{x}}\, dx$ may be written

$\int \dfrac{x^2}{\sqrt{x}}\, dx + 3 \int \dfrac{x\, dx}{\sqrt{x}} + 2 \int \dfrac{dx}{\sqrt{x}}$ by (30)—i and ii

or

$\int x^{3/2}\, dx + 3 \int x^{1/2}\, dx + 2 \int x^{-1/2}\, dx$ which equals by (30)—iii

$$\dfrac{x^{5/2}}{\frac{5}{2}} + \dfrac{3\, x^{3/2}}{\frac{3}{2}} + \dfrac{2\, x^{1/2}}{\frac{1}{2}} = \dfrac{2}{5}\, x^{5/2} + 2\, x^{3/2} + 4\, x^{1/2}.$$

The result of integration can always be tested by differentiation.

Thus using the result of Example 1 we have:

$$\dfrac{d\left(\frac{2}{5}\, x^{5/2} + 2\, x^{3/2} + 4\, x^{1/2}\right)}{dx} = \dfrac{2}{5} \cdot \dfrac{5}{2}\, x^{3/2} + 2 \cdot \dfrac{3}{2}\, x^{1/2} + 4 \cdot \dfrac{1}{2}\, x^{-1/2}$$

$$= x^{3/2} + 3\, x^{1/2} + \dfrac{2}{x^{1/2}} = \dfrac{x^2 + 3x + 2}{\sqrt{x}}$$

Example 2. $\int \sin^2 \dfrac{\theta}{2}\, d\theta$

We know that $2 \sin^2 \dfrac{\theta}{2} = 1 - \cos \theta$

or

$$\sin^2 \dfrac{\theta}{2} = \dfrac{1}{2} - \dfrac{1}{2} \cos \theta$$

Therefore

$$\int \sin^2 \frac{\theta}{2}\, d\,\theta = \int \left(\frac{1}{2} - \frac{1}{2}\cos\,\theta\right) d\,\theta = \frac{1}{2}\int d\,\theta$$

$$-\frac{1}{2}\int \cos\,\theta\, d\,\theta \quad (30)\text{—i and ii}$$

$$= \frac{1}{2}\,\theta - \frac{1}{2}\sin\,\theta \quad (30)\text{—v}$$

Example 3.

$$\int \sqrt{x}\,(x-2)^2\, dx = \int x^{1/2}\,(x^2 - 4\,x + 4)\, dx =$$

$$\int (x^{5/2} - 4\,x^{3/2} + 4\,x^{1/2})\, dx = \int x^{5/2}\, dx - 4\int x^{3/2}\, dx + 4\int x^{1/2}\, dx$$

$$= \frac{x^{7/2}}{\frac{7}{2}} - \frac{4\,x^{5/2}}{\frac{5}{2}} + \frac{4\,x^{3/2}}{\frac{3}{2}} = \frac{2}{7}\,x^{7/2} - \frac{8}{5}\,x^{5/2} + \frac{8}{3}\,x^{3/2}.$$

The third method of integration is called *substitution of a new variable.* It consists in choosing a new variable, related to the original variable in some certain way, so that when we substitute and obtain the integrand in terms of the new variable we can recognize it by the first method. Every method of integration in the end reduces to the recognition of fundamental forms such as those given in (30).

Example 1. $\int \sqrt{x^2 + 1}\,.\,x\, dx$

Put $x^2 + 1 = u$; then $2\,x\, dx = du$, $x\, dx = \dfrac{du}{2}$ and we may write

$$\int \sqrt{x^2 + 1}\,x\, dx = \int \sqrt{u}\,.\,\frac{du}{2} = \frac{1}{2}\int u^{1/2}\, du = \frac{1}{2}\,.\,\frac{u^{3/2}}{\frac{3}{2}}$$

$$= \frac{1}{3}\,u^{3/2} = \frac{1}{3}\,(x^2 + 1)^{3/2}$$

Or we may put $\sqrt{x^2 + 1} = u$. $x^2 + 1 = u^2$, $2 x\, dx = 2 u\, du$, $x\, dx = u\, du$. Then

$$\int \sqrt{x^2 + 1}\, x\, dx = \int u \cdot u\, du = \int u^2 du = \frac{u^3}{3} = \tfrac{1}{3}\, (x^2 + 1)^{3/2}$$

the same result as above.

Example 2. $\displaystyle\int \sin 2 x\, dx.$

Put $\quad\quad 2 x = u,\ 2\, dx = du,\ dx = \dfrac{du}{2}.$ $\quad\quad$ Then

$$\int \sin 2 x\, dx = \int \sin u \cdot \frac{du}{2} = \frac{1}{2} \int \sin u\, du = -\frac{1}{2} \cos u$$

$$= -\frac{1}{2} \cos 2\, x.$$

There are many rules for the choice of a new variable for special forms of functions. We shall give only one general rule.

If one part of the integrand is (except, perhaps, for a constant factor) the differential of another part, put that *other part* equal to a new variable and substitute.   (31)

In general it may be said if recognition and transformation fail, try some substitution.

Example 1. $\displaystyle\int (x^{3/2} + 5)\, x^{1/2}\, dx.$

Here $x^{1/2}\, dx$ is the differential (constant factor excepted) of $x^{3/2} + 5$, therefore, by (31) we put

$$x^{3/2} + 5 = u,\ \frac{3}{2} x^{1/2}\, dx = du,\ x^{1/2}\, dx = \frac{2}{3}\, du.\quad\quad \text{Then}$$

$$\int (x^{3/2} + 5)\, x^{1/2}\, dx = \int u \cdot \frac{2}{3}\, du = \frac{2}{3} \int u\, du = \frac{2}{3} \cdot \frac{u^2}{2}$$

$$= \frac{1}{3}\, (x^{3/2} + 5)^2$$

Or we might proceed thus:

$$\int (x^{3/2} + 5)\, x^{1/2}\, dx = \int (x^2 + 5\, x^{1/2})\, dx = \int x^2 dx + 5 \int x^{1/2}\, dx$$

$$= \frac{x^3}{3} + 5 \cdot \frac{x^{3/2}}{\frac{3}{2}} = \frac{x^3}{3} + \frac{10}{3} x^{3/2}$$

The results by the two methods do not seem to agree, but if we expand the former we have

$$\frac{1}{3}\,(x^{3/2} + 5)^2 = \frac{1}{3}\,(x^3 + 10\, x^{3/2} + 25) = \frac{x^3}{3} + \frac{10}{3}\, x^{3/2} + \frac{25}{3}$$

and the results differ only by the constant $\frac{25}{3}$. Since the constant of integration, though not written, is understood in every case, both results are correct. Two results that differ by a constant term are both correct.

Example 2. $\int \sin x \cos x\, dx$

Put $\cos x = u, \quad - \sin x\, dx = du;$ then

$$\int \sin x \cos x\, dx = - \int u\, du = - \frac{u^2}{2} = - \frac{\cos^2 x}{2}$$

or, put $\sin x = u, \quad \cos x\, dx = du$

$$\int \sin x \cos x\, dx = \int u\, du = \frac{u^2}{2} = \frac{\sin^2 x}{2}$$

But $\sin^2 x = 1 - \cos^2 x;$ therefore

$\frac{\sin^2 x}{2} = \frac{1}{2} - \frac{1}{2} \cos^2 x$ which differs from the first result by

the constant $\frac{1}{2}$. Or, again, since $2 \sin x \cos x = \sin 2 x$

$$\int \sin x \cos x\, dx = \frac{1}{2} \int \sin 2 x\, dx$$

Put $\qquad 2x = u, \;\; 2\,dx = du, \;\; dx = \dfrac{1}{2}\,du \qquad$ and

$$\int \sin x \cos x \, dx = \frac{1}{2}\int \sin 2x \, dx = \frac{1}{4}\int \sin u \, du$$

$$= -\frac{1}{4}\cos u = -\frac{1}{4}\cos 2x = -\frac{1}{4}(\cos^2 x - \sin^2 x)$$

$$= -\frac{1}{4}(\cos^2 x - 1 + \cos^2 x) = -\frac{1}{2}\cos^2 x + \frac{1}{4}$$

The three forms of the result are equally correct. Also it will be noted (last method) that we sometimes use both a transformation and a substitution before we recognize the integral.

Example 3. Find the function, y, of which $x^2 + 3x$ is the derivative, given that $y = 6$ when $x = 1$.

$$y = \int (x^2 + 3x)\, dx = \frac{x^3}{3} + \frac{3x^2}{2} + c$$

But $\qquad\qquad y = 6,\; x = 1, \qquad\qquad$ therefore

$$6 = \frac{1}{3} + \frac{3}{2} + c \text{ or } c = \frac{25}{6} \qquad\qquad \text{and}$$

$$y = \frac{x^3}{3} + \frac{3x^2}{2} + \frac{25}{6}$$

Example 4. The slope of a curve which passes through $x = 1,\; y = 3$ is $\dfrac{dy}{dx} = \dfrac{x}{y}$. Find the equation of the curve.
We have

$$\frac{dy}{dx} = \frac{x}{y} \text{ or } y\,dy = x\,dx$$

therefore $\qquad\qquad \displaystyle\int y\,dy = \int x\,dx$

or $\qquad\qquad\qquad \dfrac{y^2}{2} = \dfrac{x^2}{2} + c$

But $x = 1$, $y = 3$; therefore, $\dfrac{9}{2} = \dfrac{1}{2} + c$ or $c = 4$ and the equation is

$$\frac{y^2}{2} = \frac{x^2}{2} + 4 \text{ or } y^2 - x^2 = 8.$$

Example 5. The velocity at any time, t, of a particle moving in a straight line is $v = 5 - 3\,t$. Express the distance as a function of the time, given that $s = 15$ when $t = 2$.

We have $\dfrac{ds}{dt} = 5 - 3\,t$ or $ds = (5 - 3\,t)\,dt$

Therefore $s = \displaystyle\int (5 - 3\,t)\,dt = 5\,t - \dfrac{3\,t^2}{2} + c$

But $s = 15$, $t = 2$, therefore $15 = 10 - 6 + c$, $c = 11$ and the equation is

$$s = 5\,t - \frac{3\,t^2}{2} + 11.$$

It should be noted, in connection with this problem, that the constant of integration ($c = 11$) denotes the position of the particle when it begins to move; that is, the value of s when $t = 0$.

EXAMPLES

Integrate the following functions and prove your result correct.

1. $\displaystyle\int (1 + \sqrt{x})\,dx$ 2. $\displaystyle\int (\sqrt{x} + x^2)\,dx$ 3. $\displaystyle\int \left(x - \frac{1}{\sqrt{x}}\right)dx$

4. $\displaystyle\int (x^{1/3} - x^{-1/3})\,dx$ 5. $\displaystyle\int (x + x^{1/4})\,dx$ 6. $\displaystyle\int (\sqrt{x} + \sqrt[3]{x})\,dx$

Integrate each of the following functions by two methods and show that the results differ only by the constant of integration.

7. $\displaystyle\int (1 - x)^3\,dx$ 8. $\displaystyle\int (2 + x)^2\,dx$ 9. $\displaystyle\int (x - 3)^2\,dx$

10. $\displaystyle\int (u + 4)^3\,du$ 11. $\displaystyle\int (s - 5)^2\,ds$ 12. $\displaystyle\int (\theta - \sqrt{2})^3\,d\theta$

Integrate the following functions:

13. $\int (1 + \sqrt{x})^2 \, dx$

14. $\int \left(1 - \frac{1}{x^{1/3}}\right)^2 dx$

15. $\int \left(1 - \frac{1}{x^2}\right)^2 dx$

16. $\int \left(z^2 + \frac{1}{z^2}\right)^2 dz$

17. $\int (y^{1/3} - y^{-1/3})^2 \, dy$

18. $\int (\sqrt{s} + \sqrt[3]{s})^2 \, ds$

19. $\int (3 - 5x)^{-7} \, dx$

20. $\int \frac{5 \, dx}{(1 + 2x)^2}$

21. $\int \sqrt{4 - 3t} \, dt$

22. $\int \sqrt{2u + 1} \, du$

23. $\int \frac{d\theta}{\sqrt{1 + 4\theta}}$

24. $\int \frac{\cdot dt}{\sqrt{2 - 3t}}$

25. $\int x \, (x^2 + 1)^3 \, dx$

26. $\int t^2 \, (2 - t^3)^2 \, dt$

27. $\int u^2 \, (3 + u^3)^3 \, du$

28. $\int \frac{x^2 \, dx}{(3 + 2x^3)^3}$

29. $\int \frac{s^3 ds}{(5 - s^4)^3}$

30. $\int \frac{x^4 \, dx}{(2 - x^5)^{-1}}$

31. $\int \frac{s^{2/3} \, ds}{(2s^{5/3} + 5)^{1/2}}$

32. $\int \frac{y^{3/4} \, dy}{(1 - y^{7/4})^{2/3}}$

33. $\int \frac{x^{1/2} \, dx}{(2x^{3/2} + 1)^{1/4}}$

34. $\int (x + 1) \sqrt{x^2 + 2x} \, dx$

35. $\int \frac{(x + 1) \, dx}{\sqrt{x^2 + 2x}}$

36. $\int \sqrt{3x^2 - 6x} \cdot (x - 1) \, dx$

37. $\int \sin^2 \frac{x}{2} \cos \frac{x}{2} \, dx$

38. $\int \frac{\cos 2u \, du}{\sin^2 2u}$

39. $\int \frac{\sin t \, dt}{\sqrt{\cos t}}$

40. $\int \cos^3 3x \sin 3x \, dx$

41. $\int \frac{\sin 3x \, dx}{\cos^3 3x \, dx}$

42. $\int \cos 4x \sin 4x \, dx$

In the following examples find the function, y, of which the given expression is the derivative under the conditions named.

43. $2x^2 - x$; $y = 5$, $x = 0$

47. $5t + 4$; $y = 3$, $t = 3$

44. $x - x^3$; $y = 2$, $x = 1$

48. $4 - 5t$; $y = 4$, $t = 2$

45. $x + \sqrt{x}$; $y = 0$, $x = 4$

49. $32t + b$; $y = 5$, $t = 1$
$y = 16$, $t = 2$

46. $x^2 + x + 2$; $y = 1$, $x = 1$

50. $32t - b$; $y = 4$, $t = 1$
$y = 20$, $t = 3$

51. The slope of a curve is given by $\dfrac{dy}{dx} = 3 + 4x$. Find the equation of the curve if it passes through the point $x = 1$, $y = 2$.

52. The slope of a curve is given by $\dfrac{dy}{dx} = 2x - 5$. Find the equation of the curve if it passes through the point $x = 2$, $y = 2$.

53. A body moves in a straight line so that its velocity at any time t is given by $v = 16 - 4t$. Express the distance, s, as a function of t, given that $s = 20$ when $t = 5$.

54. A body moves in a straight line so that its velocity at any time t is given by $v = 16 + 3t$. Express the distance, s, as a function of t, given that $s = 100$ when $t = 2$.

55. A body moves in a straight line so that its velocity at any time t is given by $v = 16 + bt$. Express the distance, s, as a function of t, given that $s = 10$ when $t = 1$ and $s = 100$ when $t = 5$.

56. A body moves in a straight line so that its velocity at any time t is given by $v = 16 - bt$. Express the distance, s, as a function of t, given that $s = 0$ when $t = 1$ and $s = 20$ when $t = 3$.

57. A particle rolls down a smooth inclined plane so that $v = 14.2\,t$. Express the distance it rolls (s) as a function of t. How far will it roll in the time between $t = 0$ and $t = 2$? In the second and third seconds, the unit of time being the second, the unit of distance the foot?

Find the function of which each of the following expressions is the derivative, and sketch the graph of the function when the value of the constant of integration is $c = 0$, $c = 1$, $c = -2$.

58. 3

59. $3x - 2$

60. $3 - x - x^2$

61. $2x$

62. $x^2 - 2x + 3$

63. $4x^3$

64. $v = 4$

65. $v = 2t + 3$

66. $v = 3t^2 + 2t - 3$

67. $v = 4 - t$

68. $v = 3t^2$

69. $v = \dfrac{4}{t^2}$

In examples 64–69 v is velocity, t is time.

CHAPTER VI

INTEGRATION. APPROXIMATE SUMMATION. THE DEFINITE INTEGRAL

19. Approximate Summation. Let us consider the function $y = x^2$, of which the graph is given in Fig. 19, and ask,

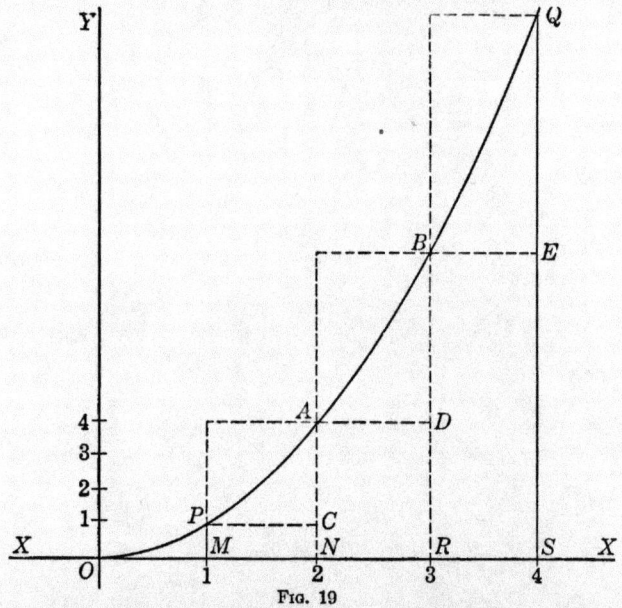

Fig. 19

what is the area bounded by the curve, the axis of x and the two ordinates corresponding to $x = 1$ and $x = 4$; in the figure the area of $PMSQ$? To obtain this area we proceed in a similar manner to that used in the elementary geometry to obtain the area of a circle. We divide the line MS into any convenient number of parts—three in the figure—and at the points of division erect ordinates to the curve, MP, NA, RB and SQ. At the points where these ordinates touch the curve we draw PC, AD and BE parallel to the axis of x. We thus have three rectangles constructed, inscribed in the curve, and can compute the areas of the rectangles. Thus

$$PMNC = MP \times MN = 1 \times 1 = 1$$
$$ANRD = NA \times NR = 4 \times 1 = 4$$
$$BRSE = RB \times RS = 9 \times 1 = 9$$

and the sum of the rectangles equals $1 + 4 + 9 = 14$, which is obviously smaller than the area bounded by the curve by the sum of the triangular pieces PCA, ADB,

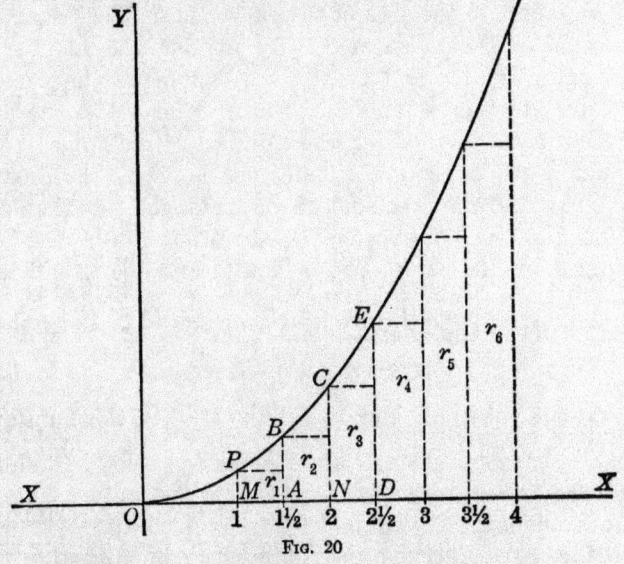

Fig. 20

etc., left over. Let us next double the number of rectangles by bisecting each of the lines MN, NR, etc., and constructing the new set of inscribed rectangles, as shown in Fig. 20. The area would now be, since

$$r_1 = \text{area } PA = MP \times MA = 1 \times \tfrac{1}{2} = \tfrac{1}{2}$$

$$r_2 = \text{area } BN = AB \times AN = (1 + \tfrac{1}{2})^2 \times \tfrac{1}{2} = \tfrac{9}{8} *$$

$$r_3 = \qquad\qquad\qquad (2)^2 \times \tfrac{1}{2} = 2$$

$$r_4 = \qquad\qquad\qquad (2 + \tfrac{1}{2})^2 \times \tfrac{1}{2} = \tfrac{25}{8}$$

$$r_5 = \qquad\qquad\qquad (3)^2 \times \tfrac{1}{2} = \tfrac{9}{2}$$

$$r_6 = \qquad\qquad\qquad (3 + \tfrac{1}{2})^2 \times \tfrac{1}{2} = \tfrac{49}{8}$$

and the sum of the rectangles is

$$\Sigma r \; \dagger = \tfrac{1}{2} + \tfrac{9}{8} + 2 + \tfrac{25}{8} + \tfrac{9}{2} + \tfrac{49}{8} = \tfrac{139}{8} = 17\tfrac{3}{8} = 17.4$$

Obviously this result is nearer than the first result (14 square units) to the area of the curve, as smaller triangular pieces have been neglected. By making the number of rectangles greater and greater we can make the sum of the areas of the rectangles nearer and nearer to the area of the curve. By taking the number of rectangles great enough we can make the difference between the area of the curve and the area-sum of the rectangles less than any assigned value, however small. In other words, the area bounded by the curve, two ordinates and the axis of x is the limit of the sum of the inscribed rectangles as the number of rectangles is indefinitely increased.‡ In symbols

$$\text{Curve-area} = \text{limit } \Sigma r \qquad\qquad (32)$$

We do not yet know how to get the limit of this sum and for the present shall content ourselves with obtaining closer and closer numerical approximations.

It is evident that we might have circumscribed rectangles (see Fig. 19) instead of inscribing them, thus getting successive approximations always a little larger than the

---

* $AB = (1 + \tfrac{1}{2})^2$, since the curve is $y = x^2$ and the ordinate is therefore always the square of the corresponding value of the abscissa.

† Read, Sigma R and meaning the sum of the r's.

‡ Compare this discussion with that in Articles 5 and following, Chap. II.

curve-area. In the first case above we should have obtained, using R to mean any circumscribed rectangle,

$$R_1 = 4 \times 1 = 4, \ R_2 = 9 \times 1 = 9, \ R_3 = 16 \times 1 = 16 \text{ and}$$
$$\Sigma R = 4 + 9 + 16 = 29$$

We have already seen that in this case

$$\Sigma r = 14$$

so that the curve-area lies between 14 and 29 square units, a very rough approximation.

Using the second series of circumscribed rectangles, where we pass along OX at intervals of one-half (that is, $\Delta x = \frac{1}{2}$) we have

$$R_1 = (1 + \tfrac{1}{2})^2 (\tfrac{1}{2}) = \tfrac{9}{8}$$
$$R_2 = (2)^2 (\tfrac{1}{2}) = 2$$
$$R_3 = (2 + \tfrac{1}{2})^2 (\tfrac{1}{2}) = \tfrac{25}{8}$$
$$R_4 = (3)^2 (\tfrac{1}{2}) = \tfrac{9}{2}$$
$$R_5 = (3 + \tfrac{1}{2})^2 (\tfrac{1}{2}) = \tfrac{49}{8}$$
$$R_6 = (4)^2 (\tfrac{1}{2}) = 8$$

and

$$\Sigma R = \tfrac{9}{8} + 2 + \tfrac{25}{8} + \tfrac{9}{2} + \tfrac{49}{8} + 8$$
$$= \tfrac{199}{8} = 24\tfrac{7}{8} = 24.9$$

We have already seen that in this case

$$\Sigma r = 17\tfrac{3}{8} = 17.4$$

so that the curve-area lies between 17.4 and 24.9 square units, a closer approximation.

A closer approximation still, and more quickly obtained, is got by taking the half sum of the inscribed and circumscribed rectangles. This is equivalent to drawing straight lines PA, AB, BQ, etc. (Fig. 19) and taking the sum of the trapezoids $APMN$, etc. Thus in the above problem:

Case 1. Curve-area $= \frac{1}{2}(\Sigma r + \Sigma R) = \frac{1}{2}(14 + 29) = 21.5$

Case 2. Curve-area $= \frac{1}{2}(\Sigma r + \Sigma R) = \frac{1}{2}(17.4 + 24.9) =$
    21.1

84 CALCULUS AND GRAPHS

We shall find later that the exact area in this case is

$$\text{Curve-area} = \text{limit } \Sigma r = \text{limit } \Sigma R = 21.$$

The subject of approximate summation is so important that we shall work another illustrative example.

Example. Find approximately the area lying above OX bounded by the curve $y = \sqrt{x}$, the axis of x and the two ordinates corresponding to $x = \frac{1}{2}$ and $x = 4$, taking intervals along OX of $\Delta x = \frac{1}{2}$. Fig. 21.

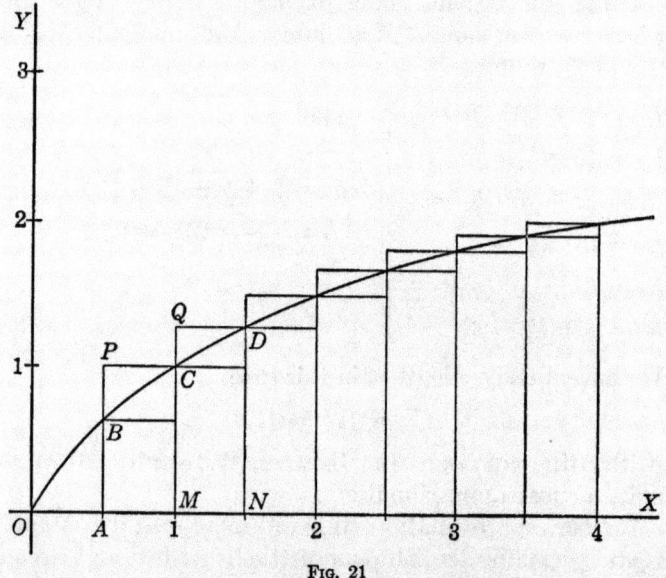

Fig. 21

We have, proceeding as in the first example, since $y = \sqrt{x}$,

$AB = \sqrt{\frac{1}{2}}$, $MC = AP = \sqrt{1}$, etc., and $OA = AM =$ etc. $= \Delta x = \frac{1}{2}$.

$$\Sigma r = (\sqrt{\tfrac{1}{2}})\,(\tfrac{1}{2}) + (\sqrt{1})\,(\tfrac{1}{2}) + (\sqrt{1 + \tfrac{1}{2}})\,(\tfrac{1}{2}) + (\sqrt{2})\,(\tfrac{1}{2})$$
$$+ (\sqrt{2 + \tfrac{1}{2}})\,(\tfrac{1}{2}) + (\sqrt{3})\,(\tfrac{1}{2}) + (\sqrt{3 + \tfrac{1}{2}})\,(\tfrac{1}{2})$$

$$= \tfrac{1}{2}\{\sqrt{\tfrac{1}{2}} + \sqrt{1} + \sqrt{1+\tfrac{1}{2}} + \sqrt{2} + \sqrt{2+\tfrac{1}{2}} + \sqrt{3} \\ + \sqrt{3+\tfrac{1}{2}}\}$$

and

$$\Sigma R = (\sqrt{1})\,(\tfrac{1}{2}) + (\sqrt{1+\tfrac{1}{2}})\,(\tfrac{1}{2}) + (\sqrt{2})\,(\tfrac{1}{2}) + (\sqrt{2+\tfrac{1}{2}})\,(\tfrac{1}{2}) \\ + (\sqrt{3})\,(\tfrac{1}{2}) + (\sqrt{3+\tfrac{1}{2}})\,(\tfrac{1}{2}) + (\sqrt{4})\,(\tfrac{1}{2})$$

$$= \tfrac{1}{2}\{\sqrt{1} + \sqrt{1+\tfrac{1}{2}} + \sqrt{2} + \sqrt{2+\tfrac{1}{2}} + \sqrt{3} \\ + \sqrt{3+\tfrac{1}{2}} + \sqrt{4}\}$$

Then

$$\Sigma r + \Sigma R = \tfrac{1}{2}\{\sqrt{\tfrac{1}{2}} + 2\sqrt{1} + 2\sqrt{1+\tfrac{1}{2}} + 2\sqrt{2} + 2\sqrt{2+\tfrac{1}{2}} \\ + 2\sqrt{3} + 2\sqrt{3+\tfrac{1}{2}} + \sqrt{4}\}$$

$$= \tfrac{1}{2}\{.71 + 2 + 2.44 + 2.82 + 3.16 + 3.46 \\ + 3.74 + 2\}$$

$$= \tfrac{1}{2}\{20.33\}$$

and

$$\text{curve-area} = \tfrac{1}{2}\,(\Sigma r + \Sigma R) = \tfrac{1}{4}\,(20.33) = 5.08$$

The exact area to the second decimal place in this case, as will be seen later, is

$$\text{Curve-area} = \text{limit } \Sigma r = \text{limit } \Sigma R = \frac{16}{3} - \frac{\sqrt{2}}{6} = 5.09$$

EXAMPLES

In the following examples sketch the graph and find the approximate curve-area between the limits named at the intervals $\Delta x = .5$ and $\Delta x = .2$.

1. $y = x^2 + 2$.   From $x = 0$ to $x = 3$.

2. $y = x^2 - 9$.   From $x = 0$ to $x = 2$.

3. $y = 4 - x^2$.   From $x = -2$ to $x = +2$.

4. $y = 9 - x^2$.   From $x = -3$ to $x = +3$.

5. $4y = 3 + 2x - x^2.$ From $x = 0$ to $x = 2$.

6. $4y = 3 + 2x - x^2.$ From $x = -1$ to $x = 3$.

7. $9y = 5 + 4x - x^2.$ From $x = 0$ to $x = 2$.

8. $9y = 5 + 4x - x^2.$ From $x = -1$ to $x = 5$.

9. $y = 2x^3.$ From $x = 1$ to $x = 3$.

10. $y = 3x^3.$ From $x = 1$ to $x = 3$.

11. $y = \sin x.$ From $x = \frac{\pi}{4}$ to $x = \frac{3\pi}{4}$; $\Delta x = \frac{\pi}{8}$.

12. $y = \sin x.$ From $x = 0$ to $x = \pi$; $\Delta x = \frac{\pi}{8}$.

13. $y = \cos x$ From $x = -\frac{\pi}{4}$ to $x = \frac{\pi}{4}$; $\Delta x = \frac{\pi}{8}$.

14. $y = \cos x.$ From $x = 0$ to $x = \pi$; $\Delta x = \frac{\pi}{8}$.

15. The depth of the water on the face of a vertical dam is measured at intervals of one foot and found to be

Distance from bank	0	1	2	3	4	5	6	7	8	9
Depth	$2\frac{1}{2}$	5	$8\frac{1}{2}$	$8\frac{1}{2}$	10	$12\frac{1}{2}$	14	$11\frac{1}{2}$	15	14
Distance from bank	10	11	12							
Depth	10	$6\frac{1}{2}$	5							

Find approximately the area of the face of the dam.

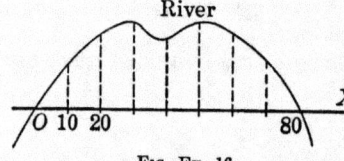

FIG. EX. 16

16. From a straight road OX distances are measured, at intervals of 10 feet from 0 to 80 feet, to the bank of a stream (see Fig.) as follows:

Dist. along road	0	10	20	30	40	50	60	70	80
Dist. to bank	0	15	22	22	20	25	22	15	0

Find approximately the area between the river and the road.

17. If the river of Example 16 has for its equation $200y = 240x - 3x^2$, the road being the axis of x as shown in the figure, find approximately the area between the river and the road, using the same intervals as in Example 16.

18. The velocity of a moving body at any time is given by the function $v = 4 - t^2$. Find approximately the distance traveled in the time from $t = 0$ to $t = 2$ at intervals $\Delta t = .2$. (Note, $s = vt$ if v is constant.)

19. The velocity of a moving body at any time is given by the function $v = 16 - t^2$. Find approximately the distance traveled in the time from $t = 0$ to $t = 4$ at intervals of $\Delta t = .5$. (See note to Example 18.)

20. If the river of Example 16 has for its equation $60\,y = 80\,x - x^2$, as in Example 17, find the approximate area between river and road. Why is this result nearer to that of Example 16 than the result of Example 17? (Suggestion: Compare the three graphs.)

CHAPTER VII

INTEGRATION AS A SUMMATION. THE DEFINITE INTEGRAL. APPLICATIONS OF THE DEFINITE INTEGRAL

20. Integration as a Summation. The Definite Integral. Let us return to the simple function $y = x^2$ and try to generalize the process discussed in Art. 19. Instead of passing from $x = 1$ to $x = 4$ at intervals of 1 or $\frac{1}{2}$, let us pass at any interval Δx. We may write for the sum of the rectangles

$$\Sigma r = (1)^2 \Delta x + (1 + \Delta x)^2 \Delta x + (1 + 2 \Delta x)^2 \Delta x + \ldots \\ + (4 - \Delta x)^2 \Delta x$$

or, adopting the sigma notation,

$\Sigma r = \sum\limits_{x=1}^{x=4-\Delta x} x^2 \Delta x$, this last being simply a condensed way of

writing the sum of the terms of the series. Then as we have seen (32) the

$$\text{Curve-area} = \text{limit } \Sigma r = \mathop{\text{limit}}\limits_{\Delta x = 0} \sum\limits_{x=1}^{x=4-\Delta x} x^2 \Delta x \tag{33}$$

If we can find the limit of this sum we can find the curve-area. If we generalize again and begin with any value of x, say $x = a$, and end with any other value, say $x = b$, instead of $x = 1$ and $x = 4$, we may write

$$\text{Curve-area} = \mathop{\text{limit}}\limits_{\Delta x = 0} \sum\limits_{x=a}^{x=b-\Delta x} x^2 \Delta x$$

Thus the problem before us is to find

$$\lim_{\Delta x = 0} \sum_{x=a}^{x=b-\Delta x} x^2 \Delta x = \lim_{\Delta x = 0} \left\{ a^2 \Delta x + (a + \Delta x)^2 \Delta x \right.$$

$$\left. + (a + 2\Delta x)^2 \Delta x + \ldots\ldots + (b - \Delta x)^2 \Delta x \right\} (34)$$

If we use dx instead of Δx to indicate that Δx approaches the limit zero and use the old-fashioned long S, $\int$, instead of limit Σ, we may write the above

$$\int_{x=a}^{x=b} x^2 dx = a^2 dx + (a + dx)^2 dx + (a + 2 dx)^2 dx +$$

$$\ldots\ldots + (b - dx)^2 dx \qquad (35)$$

We note that every term of the series is of the form $x^2 dx$ and that $x^2 dx = d\left(\dfrac{x^3}{3}\right)$.

Each term of the series (34) is of the form $x^2 \Delta x$, and also, as we note,

$$\Delta\left(\frac{x^3}{3}\right) = \frac{(x + \Delta x)^3}{3} - \frac{x^3}{3} \qquad (a)$$

$$= x^2 \Delta x + x \overline{\Delta x}^2 + \frac{\overline{\Delta x}^3}{3}$$

or

$$x^2 \Delta x = \Delta\left(\frac{x^3}{3}\right) - x \overline{\Delta x}^2 - \frac{\overline{\Delta x}^3}{3} \qquad (b)$$

Now when Δx is very small (approaches 0) then $\overline{\Delta x}^2$ is very small even as compared with Δx, and $\overline{\Delta x}^3$ is even smaller. We express this by saying that $\overline{\Delta x}^2$ and $\overline{\Delta x}^3$ are infinitesimal even as compared with Δx, or that they are infinitesimals of higher order. The equations (b and a) above may then be written, using the notation $d\left(\dfrac{x^3}{3}\right)$ to indicate that we shall neglect the terms containing $\overline{\Delta x}^2$ and $\overline{\Delta x}^3$ as being so very small; * as being, that is, infinitesimals of higher order,

$$x^2 dx = d\left(\frac{x^3}{3}\right)$$

and

$$x^2 dx = \frac{(x + dx)^3}{3} - \frac{x^3}{3} \tag{36}$$

Thus each term of the series (35) may be written as the difference of two terms (36) and the sum of the series of terms (35) as the difference of the sums of two series (36) as follows:

$$a^2 dx = \frac{(a + dx)^3}{3} - \frac{a^3}{3}$$

$$(a + dx)^2 dx = \frac{(a + 2\, dx)^3}{3} - \frac{(a + dx)^3}{3}$$

$$(a + 2\, dx)^2 dx = \frac{(a + 3\, dx)^3}{3} - \frac{(a + 2\, dx)^3}{3}$$

$$\vdots \qquad\qquad \vdots \qquad\qquad \vdots$$

$$(b - dx)^2 dx = \frac{b^3}{3} - \frac{(b - dx)^3}{3}$$

Adding in columns and noting that the first term of the positive series cancels the second of the negative series, the second of the positive the third of the negative, etc. we have

$$\int_{x=a}^{x=b} x^2 dx = a^2\, dx + (a + dx)^2\, dx + \ldots + (b - dx)^2\, dx$$

$$= \frac{b^3}{3} - \frac{a^3}{3}$$

* To illustrate this let the student put $\Delta x = .00001$ or some even smaller number and determine how small are the values of $\overline{\Delta x}^2$ and $\overline{\Delta x}^3$.

and we have succeeded in summing this infinite series of infinitesimal terms.

Since, however, $x^2 \, dx = d\left(\dfrac{x^3}{3}\right)$ we see that

$$\int x^2 dx = \int d\left(\frac{x^3}{3}\right) = \frac{x^3}{3}$$

where the symbol, $\int$, is the sign of integration. Therefore, we may write

$$\int_{x=a}^{x=b} x^2 \, dx = \left[\int x^2 \, dx\right]_a^b = \frac{x^3}{3}\bigg]_a^b = \frac{b^3}{3} - \frac{a^3}{3} \qquad (37)$$

$\int$ meaning sum; $\int$ meaning integral.

and since the symbol $\int$ meaning sum is related thus to the symbol $\int$ meaning integration we call the expression $\int_a^b x^2 \, dx$ a *definite integral* and the expression $\int x^2 \, dx$ an indefinite integral. The initial and terminal values of x (a and b) are called the lower and upper limits respectively and (37) may be expressed

To find the definite integral from $x = a$ to $x = b$ of $x^2 \, dx$, we find the indefinite integral of $x^2 \, dx$, in it substitute first the upper limit, then the lower limit and subtract the latter result from the former (38)

Note: The student may say: It is true that the terms neglected in the preceding summation are very small, but we have a very large number of them. Why do they not add up to something that can not be neglected?

But he must remember that not only is each term very small, but also that it is very small even as compared with the term just before it. Beginning with the end term, each term can be neglected in comparison with the term just before it; and, therefore, the sum of all these infinitesimal terms can be neglected in comparison with the sum as obtained by integration.

To generalize once more consider the function $y = f(x)$. Then

$$\int_a^b f(x)dx = f(a)dx + f(a+dx)\,dx + \ldots + f(b-dx)\,dx \quad (39)$$

Let $\quad f(x)dx = d\,F(x) = F(x+dx) - F(x)$

Then the series of (39) may be written

$$f(a)dx = F(a+dx) - F(a)$$

$$f(a+dx)dx = F(a+2\,dx) - F(a+dx)$$

$$f(a+2\,dx)dx = F(a+3\,dx) - F(a+2\,dx)$$

$$\vdots \qquad\qquad \vdots \qquad\qquad \vdots$$

$$f(b-dx)dx \;\; = F(b) - F(b-dx)$$

Adding we obtain

$$\int_a^b f(x)dx = F(b) - F(a)$$

But since $\qquad\qquad f(x)dx = dF(x)$

$$\int f(x)dx = F(x)$$

Therefore

$$\int_a^b f(x)dx = \int f(x)dx \Big]_a^b = F(x)\Big]_a^b = F(b) - F(a) \quad (40)$$

and our theorem (38) is seen to be true for any function, provided we can find its indefinite integral.

Note: It is frequently desirable, when the method of integration by substitution is used, to carry along the limits of the new variable instead of returning to the original variable. For example

1. $\int_0^a \dfrac{x}{\sqrt{x^2 + a^2}}\, dx$. Put $x^2 + a^2 = u^2$, $x\,dx = u\,du$, and the integral

becomes $\int du$. But when $x = 0$, $u = a$; when $x = a$, $u = a\sqrt{2}$, so that we have

$$\int_0^a \frac{x\,dx}{\sqrt{x^2 + a^2}} = \int_a^{a\sqrt{2}} du = u \Big]_a^{a\sqrt{2}} = a\sqrt{2} - a.$$

2. $\int_0^4 \frac{\sqrt{x}\,dx}{(x^{3/2} - 5)^2}$. Put $x^{3/2} - 5 = u$, $x^{1/2}\,dx = \frac{2}{3}\,du$, and the integral

becomes, since $u = -5$ when $x = 0$, and $u = 3$ when $x = 4$,

$$\frac{2}{3} \int_{-5}^3 \frac{du}{u^2} = \frac{2}{3} \left\{ -\frac{1}{u} \right\}_{-5}^3 = \frac{2}{3} \left\{ -\frac{1}{3} - \frac{1}{5} \right\} = -\frac{16}{45}.$$

EXAMPLES

Evaluate the following definite integrals.

1. $\int_0^2 (x^2 + 2x - 3)\,dx$ 2. $\int_4^1 \left(x^3 - \frac{1}{x^2} \right) dx$ 3. $\int_4^5 (t^2 - 16\,t)\,dt$

4. $\int_0^\pi \sin x\,dx$ 5. $\int_{\frac{\pi}{4}}^{\frac{\pi}{2}} \cos 2x\,dx$ 6. $\int_{\frac{\pi}{4}}^{\frac{\pi}{6}} \sin 2x \cos 2x\,dx$

7. $\int_{10}^{20} \frac{dv}{v^2}$ 8. $\int_1^2 (y + y^2 + y^3)\,dy$ 9. $\int_0^1 x\sqrt{x^2 + 1}\,dx$

21. General Meaning of the Definite Integral. For convenience and simplicity of presentation we introduced the idea of the definite integral as the problem of finding a curve-area. The general idea involved in such an integral is, however, the summation of an infinite number of infinitesimal terms, regardless of the exact meaning of these terms. This is the second remarkable intellectual tool with which the calculus supplies us. (See Art. 6.) To make use of this new tool all that is necessary is to form the element of the thing to be summed; to get a differential expression representing, in terms of some independent variable, any one of the terms to be summed. This element being found and formed, the remainder of the process consists in integrating, and substituting the limits, as shown in (40). In the following articles we shall illustrate this process by taking up some of the applications of integration as a summation.

22. Areas of Curves. Let the graph of any function, $y = f(x)$, be represented in Fig. 22, and let it be required to

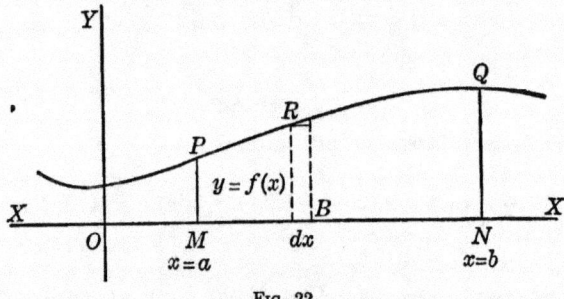

Fig. 22

find the area $PMNQ$. At any point R, between P and Q, drop ordinates to the axis of x and construct the rectangle RB as shown. When dx is taken small enough we may write the area of this rectangle, which is the element of the thing to be summed, (the term the sum of all of which will give us the area required) in the form

$$dA = y\,dx = f(x)\,dx.$$

Therefore, for the total area we have

$$A = PMNQ = \int_a^b y\,dx = \int_a^b f(x)\,dx$$

Example. Find the area of the curve $y = 1 - x^2$ from $x = -1$ to $x = 1$. We have

$$dA = y\,dx = (1 - x^2)\,dx$$

and

$$A = \int_{-1}^{1} (1 - x^2)dx = x - \frac{x^3}{3}\Bigg]_{-1}^{1} = (1 - \tfrac{1}{3}) - (-1 + \tfrac{1}{3})$$

$$= 1 - \tfrac{1}{3} + 1 - \tfrac{1}{3} = \tfrac{4}{3} = 1\tfrac{1}{3} \text{ square units.}$$

EXAMPLES

Find the curve-area for each of the following curves between the limits named and compare the results with those of the examples under Art. 19, Chapter VI.

1. $y = x^2 + 2$
   $x = 1$ to $x = 3$

2. $y = x^2 - 9$
   $x = 1, x = 2$

3. $y = 4 - x^2$
   $x = -2, x = 2$

4. $y = 9 - x^2$
   $x = -3, x = 3$

5. $4y = 3 + 2x - x^2$
   $x = 0, x = 2$

6. $4y = 3 + 2x - x^2$
   $x = -1, x = 3$

7. $9y = 5 + 4x - x^2$
   $x = 0, x = 2$

8. $9y = 5 + 4x - x^2$
   $x = -1, x = 5$

9. $y = 2x^3$
   $x = 1, x = 3$

10. $y = 3x^3$
    $x = 1, x = 3$

11. $y = \sin x$
    $x = \frac{\pi}{4}, x = \frac{3\pi}{4}$

12. $y = \sin x$
    $x = 0, x = \pi$

13. $y = \cos x$
    $x = -\frac{\pi}{4}, x = \frac{\pi}{4}$

14. $y = \cos x$
    $x = 0, x = \pi$

15. Solve problems 17 and 20, Art. 19, by integration.

16. Find the area between the two curves $y^2 = 4x$ and $x^2 = 4y$.

17. Find the area in the first quadrant between the curves $y = \sin x$, $y = \cos x$ and the axis of y.

18. Find the area between the curves $y = \sin x$, $y = \cos x$ and the axis of x from $x = 0$ to $x = \frac{\pi}{2}$.

19. Find the area enclosed by the graphs of the two functions $y = x^2$ and $y = x$.

20. Find the area between the lines $y = \sin x$ and $y = \dfrac{\sqrt{2}}{2}$, from $x = \frac{\pi}{4}$ to $x = \frac{3\pi}{4}$.

21. Find the area bounded by the lines $y = \cos x$ and $2\,y = \sqrt{2}$ and the axis of x.

22. Find the area between the curve $y = x^2$ and the line $y = 16$.

23. Find the area between the two lines $x^2y = 1$ and $3\,x + 4\,y = 7$.

23. Volumes of Surfaces of Revolution. Let the curve $y = f(x)$ be revolved about the axis of x. (See Fig. 23.)

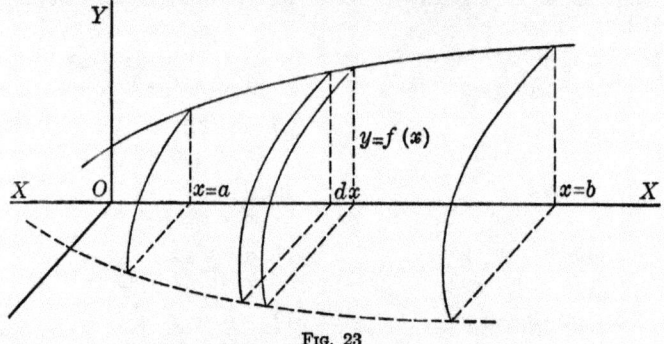

FIG. 23

The small rectangle $y\,dx$ will generate a right circular cylinder of radius y and height dx, of which the volume is $\pi\,y^2\,dx = dv$. This is the element of volume, dx being taken small enough, which being summed will give the total volume of the surface of revolution. Thus

$$V = \pi \int_a^b y^2\,dx \qquad (41)$$

If the curve is revolved about $O\,Y$ we have

$$dv = \pi\,x^2\,dy \text{ and } v = \int_{y=c}^{y=d} x^2\,dy$$

Example. Find the volume generated by revolving the curve $y = x^2$ about (a) $O\,X$; (b) $O\,Y$; (c) the line $y = -2$; from $x = 0$ to $x = 2$ (see Fig. 19).

(a) $v = \pi \int_0^2 y^2 dx = \pi \int_0^2 x^4\, dx = \pi \dfrac{x^5}{5}\Big]_0^2 = \dfrac{32\,\pi}{5}$

(b) $v = \pi \int_0^4 x^2 dy = \pi \int_0^4 y\, dy = \dfrac{\pi y^2}{2}\Big]_0^4 = 8\,\pi$

(c) In this case the radius of the circular cross section is $y + 2$.

$$v = \pi \int_0^2 (y + 2)^2 dx = \pi \int_0^2 (x^2 + 2)^2 dx$$

$$= \pi \int_0^2 (x^4 + 4\,x^2 + 4)\, dx$$

$$= \pi \left(\dfrac{x^5}{5} + \dfrac{4\,x^3}{3} + 4\,x\right)_0^2 = \dfrac{376\,\pi}{15}$$

EXAMPLES

Find the volume got by revolving each of the following curves about the axis specified, between the limits given.

1. $y = \sqrt{x} + 2$ about OX, from $x = 0$ to $x = 4$.

2. $y = x + \dfrac{1}{x}$ about OX, from $x = 1$ to $x = 3$.

3. $x = y + y^2$ about OY, from $y = 0$ to $y = 4$.

4. $x = y^{1/2} - 2$ about OY, from $y = 1$ to $y = 4$.

5. $y = \sqrt{x} + 2$ about $y = -2$ from $x = 0$ to $x = 4$.

6. $y = 2\sqrt{x} + 3$ about $y = 1$ from $x = 1$ to $x = 3$.

7. The curve whose equation is $\dfrac{x^2}{a^2} + \dfrac{y^2}{b^2} = 1$ is called an ellipse. Find the volume got by revolving the ellipse (1) about OX. (2) about OY.

8. Given that the curve $x^2 + y^2 = a^2$ is a circle with radius a, find by integration the volume of a sphere of radius a.

9. The curve $xy = 10$, called an hyperbola, is revolved about OY. Find the volume generated from $y = 1$ to $y = 4$.

10. Find the volume generated by revolving the arch of the curve $y = x^2 - 2x$ about the axis of x.

11. Find the volume generated by revolving the arch of the curve $x = 2y - y^2$ about the axis of y.

12. The equation $x^2 + y^2 = 25$ represents a circle of radius 5. Find, by revolving the circle about OX, the volume of the spherical cap (called segment) from $x = 2$ to $x = 5$.

13. The circle of example 12 is cut by the straight line $y = x$. Find the volume got by revolving about OX the area bounded by the circle, the straight line and the axis of x.

14. Find, by integration, the volume of a right circular cone of height h and angular opening $\frac{\pi}{2}$.

15. The curve $y = 2\sqrt{\sin x}$ is revolved about OX. Find the volume generated from $x = 0$ to $x = \pi$.

16. The curve $y = \tan x$ is revolved about OX. Find the volume generated from $x = 0$ to $x = \frac{\pi}{4}$.

17. The curve $x = \cos y \sqrt{\sin y}$ is revolved about OY. Find the volume generated from $y = 0$ to $y = \pi$.

18. The curve $y^2 = \sin x \cos x$ is revolved about OX. Find the volume generated by an arch of the curve.

24. Fluid Pressure.

We have learned in physics that the pressure on a surface immersed in a fluid is equal to the weight of the fluid which rests upon the surface as a base, and that the pressure is exerted equally in all directions. If, therefore, the surface be horizontal the pressure can be found by multiplying the area of the surface by the depth of the fluid, and multiplying this product by the weight of the fluid per cubic unit. But if the surface the pressure on which we wish to measure is not horizontal, the depth of the fluid resting upon it is not constant. This is, therefore, a case in which we have to form the element of the thing to be summed, and to perform the summation by integration. We accomplish this by taking a strip of the area of the immersed surface, so narrow that the whole strip can be regarded as at the same depth in the fluid. Thus, in Fig. 24, let OY be the surface of the fluid, whose

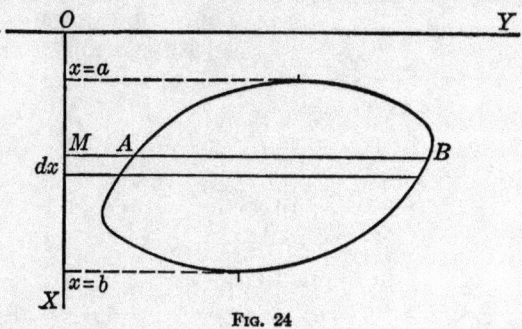

Fig. 24

density is ω, and let the positive axis of X be taken downward. Take any horizontal strip of area AB of width dx.

Let $MB = y_2$ and $MA = y_1$. Then we have

  Element of area $= dA = (y_2 - y_1) \, dx$

  Element of volume $= dV = x(y_2 - y_1) \, dx$

  Element of pressure $= dp = \omega x(y_2 - y_1) \, dx$

and the pressure

$$p = \int_{x=a}^{x=b} \omega x(y_2 - y_1) \, dx \qquad (42)$$

the values $x = a$ and $x = b$ represent the highest and lowest points, respectively, of the immersed surface.

The density, ω, is usually a constant (for water $\omega = 62.5$ lbs. per cubic foot) but it may be a variable.

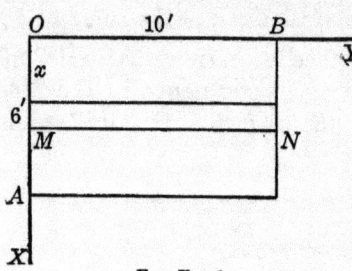

FIG. EX. 1

Example 1. What is the pressure on a rectangular flood-gate 6 feet deep and 10 feet broad if the surface of the water is level with the top of the gate? See Fig.

In this case $dA = MN\,dx$

$$= 10\,dx$$

$$dv = x\,.\,10\,dx$$

$$dp = \omega x\,10\,dx$$

and $p = \displaystyle\int_0^6 10\omega x\,dx = 10\,\omega\,\dfrac{x^2}{2}\Big]_0^6 = 11250\text{ lbs.} = 5\tfrac{5}{8}\text{ tons.}$

Example 2. Suppose the gate of Example 1 is the end of a closed sluice, and that the level of the water is 50 feet above the top of the gate; then

$$dA = 10\,dx,\quad dv = (50 + x)\,10\,dx,\quad dp = \omega\,(50 + x)\,10\,dx$$

and

$$p = \int_0^6 10\,\omega\,(50 + x)\,dx = 10\,\omega\left\{50\,x + \frac{x^2}{2}\right\}_0^6 = 99\tfrac{3}{8}\text{ tons.}$$

Example 3. Suppose the water, in Example 1, rises to within 2 feet of the top of the gate; then, as before, $dp = 10\omega x\,dx$ and

$$p = \int_2^6 10\,\omega x\,dx = 10\,\omega\left(\frac{x^2}{2}\right)_2^6 = 5\text{ tons.}$$

Example 4. Suppose the water in Example 1 is so mixed with oil discharged from a factory that the density varies as the depth below the surface, being 62.5 lbs. per cubic foot at a depth of 6 feet.   Then

$$\omega = Kx \text{ and } \omega = 62.5 \text{ when } x = 6, \text{ so that } K = \frac{62.5}{6} \text{ and}$$

$$\omega = \frac{62.5}{6} x. \text{ We thus have}$$

$$dp = \frac{62.5}{6} x \cdot x \cdot 10 \, dx$$

and

$$p = \frac{625}{6} \int_0^6 x^2 dx = \frac{625}{6} \left(\frac{x^3}{3} \right)_0^6 = 7500 \text{ lbs.} = 3\tfrac{3}{4} \text{ tons.}$$

EXAMPLES

Find the pressure on an equilateral triangle, of 10 foot side, immersed vertically in water, under the conditions named in Examples 1–4.

1. When the base is in the surface of the water.

2. When the vertex is in the surface of the water and the base horizontal.

3. When the base is horizontal and 2 feet below the surface of the water, the vertex being downwards.

4. When the triangle is in the position of Example 2 except that the vertex is 4 feet below the surface of the water.

5. A triangle with base 4 feet and altitude 5 feet is immersed vertically with its base in the surface of the water. Find the pressure on the triangle.

6. Find the pressure on the triangle of Example 5 when the triangle is inverted.

7. A trough has vertical ends in the form of a right triangle with the hypothenuse horizontal, each leg of the triangle being 14 inches. The trough contains mercury to within an inch of the top. Find the pressure on one end of the trough. (Take the specific gravity of mercury as 13; i. e., 1 cubic foot of mercury weighs 62.5 × 13 pounds.)

8. A sluice is closed by a water-gate in the form of a trapezoid, upper base 6 feet, lower base 4 feet, and with sides equally inclined to the

base. Find the pressure on the gate when the sluice is full, if the level of the pond which fills the sluice is 40 feet above the top of the gate, and the gate is 3 feet high.

9. A right triangle with sides 3, 4 and 5 feet respectively is immersed vertically with the 3 foot side in the surface of the water. What is the pressure on the triangle? To what depth must the triangle be sunk, keeping the 3 foot side parallel to the surface of the water, to double the pressure?

10. To what height must the triangle in Example 9 be raised to halve the pressure?

11. A square of side a feet is placed in water with its diagonal vertical, and a corner in the surface. Find the pressure on the square.

12. Taking the surface of the water as the axis of y and the vertical line downward as the axis of x, find the pressure on the triangle formed by the axis of y and the two lines $2x + 4y = 8$ and $3x - 4y = 12$.

13. Using the same axes as in Example 12 sketch the graph of $y^2 = 8x$, and find the pressure on the area bounded by the curve and the line $x = 8$.

14. Taking the axis of y horizontal and the axis of x vertically downward sketch the graph of the curve $y^2 + 16x = 0$. If the surface of the water is in the line $x = -3$, find the pressure on the area bounded by the curve.

15. A tank is rectangular in shape, 10 feet long, 6 feet wide and 5 feet deep. It is filled with mixed oils such that the density is proportional to the depth below the surface, the density being 15 pounds when halfway down. Find the pressure on the ends and sides of the tank.

16. A rectangular dam 100 feet wide and 30 feet deep has its top 1000 feet below the surface of the water in the reservoir which supplies the water. Find the pressure on the dam in tons.

25. Moment of Force or Turning Moment.

Moment of force, or turning moment, is the product of the force by the arm; the arm being the distance from the axis about which the force acts to the point of application of the force.

Let us suppose that the force with which we are dealing is fluid pressure and that the axis about which it acts is the surface of the fluid, then (see Fig. 24) the force acting on any little strip of area AB is the pressure on that strip. We have found this to be

$$dp = \omega\, x\, (y_2 - y_1)\, dx.$$

The moment of this force is, therefore, $dM = x\, dp = \omega\, x^2\, (y_2 - y_1)\, dx$, since the arm is x, the distance from the surface. Then we have for the moment of force about the surface

$$M = \int_{x_1}^{x_2} x\, dp = \int_{x_1}^{x_2} \omega\, x^2\, (y_2 - y_1)\, dx. \tag{43}$$

Example 1. Find the moment of force about the surface of the water of the rectangle in Example 1, Art. 24.

Here $$dp = 10\,\omega\, x\, dx$$

and $$dM = 10\,\omega\, x^2 dx$$

$$M = 10\,\omega \int_0^6 x^2\, dx = 10\,\omega\, \frac{x^3}{3}\Big]_0^6 = 45{,}000 \text{ ft. lbs.}$$

Example 2. Find the moment of force in Example 1 about the bottom of the rectangle.

In this case the arm is $6 - x$ and

$$dM = 10\,\omega\, x\, (6 - x)\, dx$$

$$M = 10\,\omega \int_0^6 (6x - x^2)\, dx = 10\,\omega \left\{ 3x^2 - \frac{x^3}{3} \right\}_0^6 = 22{,}500 \text{ ft. lbs.}$$

EXAMPLES

1. Find the moment of force about the surface in Examples 1–6 of Art. 24.

2. Find the moment of force about the top of the gate in Example 8 of Art. 24.

3. Find the moment of force about the bottom of the gate in Example 8 of Art. 24.

4. Using the triangle of Example 9, Art. 24, find the moment of force about the surface.

26. Distance Determined from Velocity, Acceleration, and Time. We have seen (Arts. 11 and 12) that the velocity of a moving body is given by $v = \dfrac{ds}{dt}$, whence $ds = v\,dt$. It follows, therefore, that the distance is given by

$$S = \int v\,dt \qquad\qquad (44)$$

Also since angular velocity is given by

$$\omega = \frac{d\theta}{dt} \text{ or } d\theta = \omega\,dt$$

we have

$$\theta = \int \omega\,dt \qquad\qquad (45)$$

Since, however, in a rotating body we know that $S = a\,\theta$, then $\dfrac{ds}{dt} = a\,\dfrac{d\theta}{dt}$. We thus have, for the distance traveled by a point on a revolving circle of radius a,

$$S = \int ad\theta = \int a\,\omega\,dt = a\int \omega\,dt \qquad (46)$$

Example 1. A body moves with a velocity $v = 3 - 2\,t$. Find the distance traveled during the first 3 seconds; during the 5th second.

Here

$$v = \frac{ds}{dt} = 3 - 2\,t, \text{ or } ds = (3 - 2\,t)\,dt$$

Therefore,

$$s_1 = \int_0^3 (3 - 2\,t)\,dt = 3\,t - t^2\Big]_0^3 = 0$$

and

$$s_2 = \int_4^5 (3 - 2\,t)\,dt = 3\,t - t^2\Big]_4^5 = -6$$

These results seem peculiar. Let us analyze them. We find s_2 to be negative. But during the 5th second $v = \dfrac{ds}{dt}$ $= 3 - 2\,t$ is negative, which means that s decreases as t increases, or that the body is moving backwards; that is, in the negative direction. The body actually moves through a distance 6, however, during the 5th second.

It is found that s_1 is zero. This does not mean that the body has not moved through any distance in the first 3 seconds, but that the *algebraic sum* of those distances, some positive and some negative, is zero. Putting

$$v = \frac{ds}{dt} = 3 - 2\,t = 0 \text{ we find } t = \tfrac{3}{2}$$

Then

t	1	$\frac{3}{2}$	2
$\dfrac{ds}{dt}$	$+$	0	$-$
s	incr.		decr.

That is, for the first $1\frac{1}{2}$ seconds s is increasing and the body moves forward. From $1\frac{1}{2}$ seconds to 3 seconds s is decreasing and the body moves backward. Therefore, taking the two periods separately, we have

$$s' = \int_0^{3/2} (3 - 2\,t)\, dt = 3\,t - t^2 \Big]_0^{3/2} = \tfrac{9}{4}$$

$$s'' = \int_{3/2}^{3} (3 - 2\,t)\, dt = 3\,t - t^2 \Big]_{3/2}^{3} = -\tfrac{9}{4}$$

and $s_1 = s' - s'' = \tfrac{9}{2} = 4\tfrac{1}{2}$, distance traveled during first 3 seconds.

If we plot the graph of $v = 3 - 2\,t$ (Fig. 25) beginning

with $t = 0$, we see that the area between OV, the line ABD, and OT represents the distance traveled in any given time, since $ds = v\,dt$ and $s = \int_{t_1}^{t_2} v\,dt$ where $\int_{t_1}^{t_2} v\,dt$ means the limit of the sum of the rectangles $v\,\Delta t$. From $t = 0$ to $t = 3$ the area is

$AOB = \frac{1}{2} OB \times OA = \frac{9}{4}$, plus $BCD = \frac{1}{2} BC \times CD = -\frac{9}{4}$
The numerical value of the total area is $\frac{9}{4} + \frac{9}{4} = 4\frac{1}{2}$.

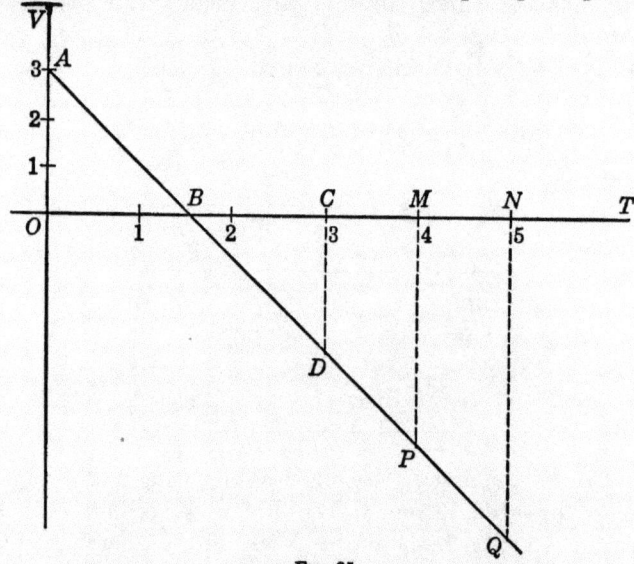

Fig. 25

Also, the trapezoid
$MNQP = \frac{1}{2}(MP + NQ) \times MN = \frac{1}{2}(-5 - 7) \times 1 = -6$ is the distance traveled from $t = 4$ to $t = 5$. This gives, graphically, the results already obtained. We may note also that when the velocity-time area is above OT the distance is positive; when below OT, the distance is negative; and that the moving particle comes to rest ($v = 0$) at the point B ($t = 1\frac{1}{2}$) where the graph crosses

the axis of T. Similarly, if $f = F(t)$, where f is acceleration, be plotted, areas as above represent velocities, since
$\frac{dv}{dt} = f$ and $dv = f\,dt$.

Example 2. A wheel revolves until friction stops it, with
angular velocity $\omega = 15 - 3\,t$. Through what angle will
the wheel turn in 2 seconds? In 5 seconds? In 6 seconds?
Here

$$\omega = \frac{d\theta}{dt} = 15 - 3\,t;\; d\theta = (15 - 3\,t)\,dt$$

and

$$\theta_1 = \int_0^2 (15 - 3\,t)\,dt = 15\,t - \frac{3\,t^2}{2}\bigg]_0^2 = 24 \text{ radians.}$$

$$\theta_2 = \int_0^5 (15 - 3\,t)\,dt = 15\,t - \frac{3\,t^2}{2}\bigg]_0^5 = 37\tfrac{1}{2} \text{ radians.}$$

$$\theta_3 = \int_0^6 (15 - 3\,t)\,dt = 15\,t - \frac{3\,t^2}{2}\bigg]_0^6 = 36 \text{ radians.}$$

The value of θ_3 is absurd because as a matter of fact the
wheel stops in 5 seconds. If the law $\omega = 15 - 3\,t$ held
after 5 seconds the motion would be reversed, and from
$t = 5$ to $t = 6$ we should find $\theta = -1\tfrac{1}{2}$. It should be
noted that these results could be got graphically as in
Example 1.

Example 3. A rifle ball is fired through a 3-inch plank,
the resistance of which causes an unknown constant retardation of its velocity. Its velocity on entering the
plank is 1000 feet a second, and on leaving the plank is
500 feet a second. How long does it take the ball to
traverse the plank? How thick must the plank be to stop
the ball?

In this case the retardation is a negative acceleration so that

$$f = \frac{d^2s}{dt^2} = \frac{dv}{dt} = - K$$

and

$$v = - Kt + c$$

But $v = 1000$ when $t = 0$, therefore, $c = 1000$ and

$$v = \frac{ds}{dt} = - Kt + 1000.$$

Also $v = 500$ when $t = T$ (the time of leaving the plank), therefore,

$$500 = -KT + 1000 \text{ or } K = \frac{500}{T} \qquad \text{(a)}$$

Again, since

$$\frac{ds}{dt} = - Kt + 1000$$

we have

$$s = \int (- Kt + 1000)\, dt = - \frac{K t^2}{2} + 1000\, t + c'$$

But $s = 0$ when $t = 0$; therefore $c' = 0$.

Also $s = 3$ inches $= \frac{1}{4}$ foot when $t = T$.

Therefore,

$$\frac{1}{4} = - \frac{KT^2}{2} + 1000\, T$$

In this substitute $K = \frac{500}{T}$ and we have

$$\frac{1}{4} = - \frac{250}{T} \cdot T^2 + 1000\, T$$

or

$$T = \frac{1}{3000} \text{ second.}$$

To answer the second question we put $v = 0$ when $t = T'$ in the general expression for v. Thus

$$0 = - KT' + 1000; \quad T' = \frac{1000}{K}$$

But since $K = \dfrac{500}{T} = \dfrac{500}{\dfrac{1}{3000}} = 1{,}500{,}000$

Therefore,

$$T' = \frac{1000}{1{,}500{,}000} = \frac{1}{1500} \text{ second.}$$

Also

$$S = - \frac{KT'^2}{2} + 1000\,T' = - \frac{1{,}500{,}000}{2} \cdot \frac{1}{(1500)^2}$$

$$+ 1000 \left(\frac{1}{1500} \right) = \frac{1}{3} \text{ foot} = 4 \text{ inches.}$$

EXAMPLES

1. If $v = 5 + 7\,t$, where v is the velocity in feet per second and t is the time in seconds, find the distance traveled (a) in the first 5 seconds, (b) during the 10th second.

2. If $v = 3\,t^2$, find the distance traveled during the 5th second.

3. If $v = 2\,t^3 + 3\,t + 5$, find the distance traveled between the ends of the 3d and 10th seconds.

4. The speed of a body at the end of t seconds from a fixed instant is given by $v = u + at$ (where u and a are constants). Show that the distance traveled in these t seconds is $ut + \frac{1}{2}\,at^2$. Show also that u is the speed at the fixed instant, and that the acceleration is constant and equal to a.

5. Find the number of revolutions made in the first 5 minutes by a wheel which moves with an angular speed $\omega = \dfrac{t^2}{1000}$ radians per second.

6. In Example 5, find the number of revolutions made during the 3d minute.

7. The velocity of a moving body is given by the equation $v = 2 \cos t$. How long and how far will the body move before it comes to rest?

8. How far will the body of Example 7 move in π seconds?

9. A boy starts a slide on ice with a speed of $14\frac{2}{3}$ feet per second. He stops in $3\frac{2}{3}$ seconds after sliding $26\frac{8}{9}$ feet. His speed at any time is given by the equation $v = -32 Kt + 14\frac{2}{3}$ where K is the coefficient of friction between the boy's shoes and the ice. Find the value of K.

10. If the boy of Example 9 starts with a speed of 20 feet per second, how long and how far will he slide before stopping?

In each of the following examples find how long and how far the body will move before it comes to rest (s in feet; t in seconds). Check your results by means of a graph.

11. $v = 5 - 4 t$.   12. $v = 4t - 5$.   13. $v = \frac{1}{2} t - 2$.

14. $f = 2 - t$; starting from rest (f = acceleration).

15. $f = 3 - 2 t$; starting from rest.

16. $f = \frac{1}{2} t - 4$; starting from rest.

How long and through what angle will a wheel turn before coming to rest if its motion is given by the following equations? Check your results by means of a graph. If the radius of the wheel is 2 feet, how far will a point on the rim of the wheel travel?

17. $\omega = 3 - 4 t$.   18. $\omega = \frac{1}{2} - t$.   19. $\omega = 2 t - 3$.

20. $a = 1 - 4 t$; starting from rest (a = angular acceleration)

21. $a = \pi - 4 t$; starting from rest.

22. $a = t - \frac{\pi}{4}$; starting from rest.

23. A torpedo, shot under water, has a speed of 50 feet per second at the moment its compressed air is exhausted. It suffers a constant retardation of 5 feet per second per second. How long and how far will it travel before coming to rest?

24. A torpedo shot under water has a speed of 60 feet per second at the moment its compressed air is exhausted. It comes to rest after traveling 350 feet. How long does it travel before coming to rest? What is the retardation?

25. A rifle bullet enters a sand bank 300 feet thick with a velocity of 1000 feet per second. If it suffers a retardation of 2000 feet per second per second, how far will it penetrate the bank before coming to rest?

26. How long will the bullet of Example 25 take to travel its whole path through the bank? The first half of the path? The second half? With what velocity does the bullet pass the midpoint of its path?

We shall conclude this portion of our work with the discussion of a few more applications of integration as a summation; cases in which we sum an infinite number of infinitesimal terms, or to express the idea a little differently, in which we find the limit of the sum of a series of very small terms, each of a certain type, as the number of terms becomes very great. Our first task in each case, as will be seen, will be, as in the preceding applications, to form the type term of the series; the element of the thing to be summed.

27. Center of Pressure and Center of Gravity. Returning to the discussion of Art. 25, let us suppose that there is a certain distance, $\bar{x}$, below the axis of y such that, if the total pressure were applied at that distance, the moment of force would be the same as that already computed. Then

$$\bar{x} \cdot \int_{x_1}^{x_2} \omega x (y_2 - y_1) \, dx = \int_{x_1}^{x_2} \omega x^2 (y_2 - y_1) \, dx$$

and

$$\bar{x} = \frac{\displaystyle\int_{x_1}^{x_2} \omega x^2 (y_2 - y_1) \, dx}{\displaystyle\int_{x_1}^{x_2} \omega x (y_2 - y_1) \, dx} \tag{47}$$

Such a distance is called the Center of Pressure of the surface. It is usually measured from the surface level of the fluid causing the pressure.

Again, if the force of which the moment is to be found, be gravity, a point can be found such that if the total force be applied at that point the turning moment will be equal to the moment of force as found by the methods of Art. 25.

Such a point is called the center of gravity of the body. Thus given the closed area bounded by $F(x, y) = 0$, where the equation may be solved for y as a function of x $\left(y = f_1(x), \text{ say}\right)$ or for x as a function of y $\left(x = f_2(y), \text{ say}\right)$, we may take an element of area parallel to OY or one parallel to OX as in Fig. 26. We thus have in the two

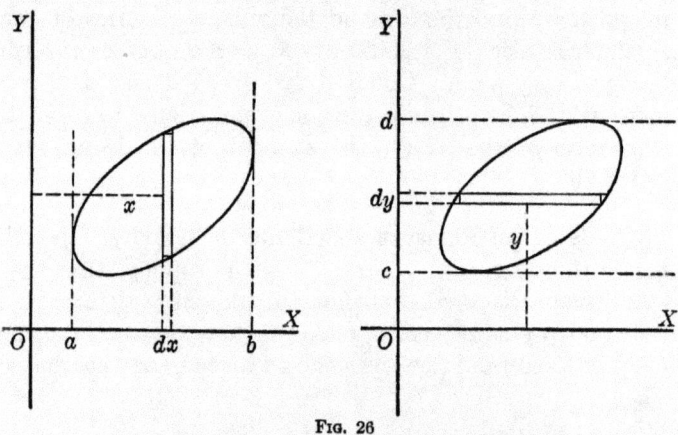

Fig. 26

cases respectively, m being the mass, ρ the density, or mass per square unit, and g the acceleration due to gravity,

$$dA = (y_2 - y_1)\, dx \qquad\qquad dA = (x_2 - x_1)\, dy$$

$$dm = \rho\, (y_2 - y_1)\, dx \qquad\qquad dm = \rho\, (x_2 - x_1)\, dy$$

$$d\text{ Force} = g\rho\, (y_2 - y_1)\, dx \qquad dF = g\rho\, (x_2 - x_1)\, dy$$

$$d\text{ Moment} = g\rho\, x\, (y_2 - y_1)\, dx \qquad dM = g\rho\, y\, (x_2 - x_1)\, dy$$

where, in the former, the force of gravity acts downward parallel to OY, and in the latter to the right parallel to OX. Therefore

$$\text{Mo. of Force} = \int_{x=a}^{x=b} g\rho\, x\, (y_2 - y_1)\, dx, \text{ or} \int_{y=c}^{y=d} g\rho\, y\, (x_2 - x_1)\, dy$$

But if the center of gravity be the point $(\overline{X}, \overline{Y})$ then, the total force being

$$F = \int dF = \int_{x=a}^{x=b} g\rho (y_2 - y_1)\, dx, \text{ or} \int_{y=c}^{y=d} g\rho (x_2 - x)\, dy,$$

we have

$$\overline{X} \int_a^b g\rho (y_2 - y_1)\, dx = \int_a^b g\rho\, x (y_2 - y_1)\, dx$$

and

$$\overline{Y} \int_c^d g\rho (x_2 - x_1)\, dy = \int_c^d g\rho\, y (x_2 - x_1)\, dy$$

or

$$\overline{X} = \frac{\displaystyle\int_a^b g\rho\, x (y_2 - y_1)\, dx}{\displaystyle\int_a^b g\rho (y_2 - y_1)\, dx} \text{ and } \overline{Y} = \frac{\displaystyle\int_c^d g\rho\, y (x_2 - x_1)\, dy}{\displaystyle\int_c^d g\rho (x_2 - x_1)\, dy} \quad (48)$$

From which the common factor g may be removed, and also ρ if the density be constant.

Example 1. Find the center of pressure of an equilateral triangle of side 4 feet, immersed vertically in water with a side in the surface of the water. See Fig.

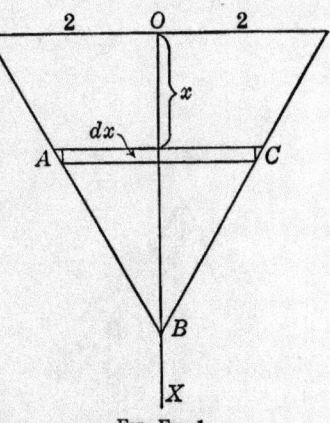

Fig. Ex. 1

Here $\qquad OB = 2\sqrt{3} \qquad$ and

$$\frac{AC}{4} = \frac{2\sqrt{3}-x}{2\sqrt{3}} \text{ or } AC = \frac{2}{\sqrt{3}}(2\sqrt{3}-x) = 4 - \frac{2x}{\sqrt{3}}$$

Therefore,

$$d\,\text{Area} = \left(4 - \frac{2x}{\sqrt{3}}\right)dx$$

$$dp = \omega x\left(4 - \frac{2x}{\sqrt{3}}\right)dx$$

$$d\,\text{Mo.} = \omega x^2\left(4 - \frac{2x}{\sqrt{3}}\right)dx$$

and

$$\overline{X}\int_0^{2\sqrt{3}} \omega x\left(4 - \frac{2x}{\sqrt{3}}\right)dx = \int_0^{2\sqrt{3}} \omega x^2\left(4 - \frac{2x}{\sqrt{3}}\right)dx$$

Whence

$$\overline{X} = \frac{\int_0^{2\sqrt{3}} x^2\left(4 - \frac{2x}{\sqrt{3}}\right)dx}{\int_0^{2\sqrt{3}} x\left(4 - \frac{2x}{\sqrt{3}}\right)dx} = \frac{\int_0^{2\sqrt{3}}\left(4x^2 - \frac{2x^3}{\sqrt{3}}\right)dx}{\int_0^{2\sqrt{3}}\left(4x - \frac{2x^2}{\sqrt{3}}\right)dx}$$

$$= \frac{\frac{4x^3}{3} - \frac{x^4}{2\sqrt{3}}\Big]_0^{2\sqrt{3}}}{2x^2 - \frac{2x^3}{3\sqrt{3}}\Big]_0^{2\sqrt{3}}} = \sqrt{3}$$

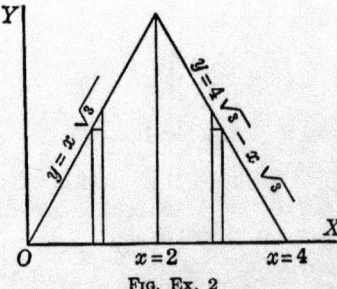

Fig. Ex. 2

Example 2. A triangle is formed by the lines $y = x\sqrt{3}$, $y = 4\sqrt{3} - x\sqrt{3}$ and the axis of x. Find the center of gravity of the triangle. See Fig.

Here $dA = y\,dx$ for each of the two right triangles formed as shown in the figure.

$$dA = x \sqrt{3} \, dx \text{ or } dA = (4\sqrt{3} - x\sqrt{3}) \, dx$$
$$dm = \rho x \sqrt{3} \, dx \text{ or } \rho (4\sqrt{3} - x\sqrt{3}) \, dx$$
$$dF = g\rho x \sqrt{3} \, dx \text{ or } g\rho (4\sqrt{3} - x\sqrt{3}) \, dx$$
$$dMo = g\rho x^2 \sqrt{3} \, dx \text{ or } g\rho x (4\sqrt{3} - x\sqrt{3}) \, dx$$

and the total moment of force is

$$M = \int_0^2 g\rho x^2 \sqrt{3} \, dx + \int_2^4 g\rho x (4\sqrt{3} - x\sqrt{3}) \, dx$$

The total force is

$$F = \int_0^2 g\rho x \sqrt{3} \, dx + \int_2^4 g\rho (4\sqrt{3} - x\sqrt{3}) \, dx$$

Therefore, assuming the density, ρ, to be constant,

$$\overline{X} \left\{ g\rho \int_0^2 x \sqrt{3} \, dx + g\rho \int_2^4 (4\sqrt{3} - x\sqrt{3}) \, dx \right\}$$
$$= g\rho \int_0^2 x^2 \sqrt{3} \, dx + g\rho \int_2^4 x (4\sqrt{3} - x\sqrt{3}) \, dx$$

or

$$\overline{X} = \frac{\int_0^2 x^2 \sqrt{3} \, dx + \int_2^4 x (4\sqrt{3} - x\sqrt{3}) \, dx}{\int_0^2 x \sqrt{3} \, dx + \int_2^4 (4\sqrt{3} - x\sqrt{3}) \, dx}$$

$$= \frac{\left[\dfrac{x^3 \sqrt{3}}{3} \right]_0^2 + \left[2\sqrt{3} \, x^2 - \dfrac{x^3 \sqrt{3}}{3} \right]_2^4}{\left[\dfrac{x^2 \sqrt{3}}{2} \right]_0^2 + \left[4\sqrt{3} \, x - \dfrac{x^2 \sqrt{3}}{2} \right]_2^4} = \frac{8\sqrt{3}}{4\sqrt{3}} = 2$$

as could have been seen from the symmetry of the figure. To find $\overline{y}$ we have

$$dA = (x_2 - x_1) \, dy = \left[\left(4 - \frac{y}{\sqrt{3}} \right) - \frac{y}{\sqrt{3}} \right] dy = \left(4 - \frac{2y}{\sqrt{3}} \right) dy$$

$$dm = \rho \left(4 - \frac{2\,y}{\sqrt{3}} \right) dy$$

$$dF = g\,\rho \left(4 - \frac{2\,y}{\sqrt{3}} \right) dy$$

$$dMo = g\,\rho\,y \left(4 - \frac{2\,y}{\sqrt{3}} \right) dy, \text{ and}$$

$$\overline{Y} = \frac{\displaystyle\int_{0}^{2\sqrt{3}} y \left(4 - \frac{2\,y}{\sqrt{3}} \right) dy}{\displaystyle\int_{0}^{2\sqrt{3}} \left(4 - \frac{2\,y}{\sqrt{3}} \right) dy} = \frac{8}{4\,\sqrt{3}} = \frac{2\,\sqrt{3}}{3}$$

EXAMPLES

1. Find the center of pressure in Examples 1–6, 8 and 9 of Art. 24.

2. Find the center of gravity of a rectangle 8 inches long and 3 inches high.

3. Find the center of gravity of a right triangle with sides 3, 4 and 5 feet, the 4-foot side being along OX.

4. Find the center of gravity of the triangle of Example 3 if the 3-foot side is along OX.

5. Find the center of gravity of the triangle of Example 3 if the hypothenuse is along OX.

6. Find the center of gravity of the area bounded by the curve $y = \sqrt{x}$ and the straight line $x = 4$.

7. Find the center of gravity of the area bounded by the curve $y = x^2$ and the line $y = 4$.

8. Find the center of gravity of the area bounded by the curve $y = x^3$, the line $x = 2$ and the axis of x.

9. An equilateral triangle, each side of length $2\,a$, has one side on the axis of x, the origin being at the middle point of that side. Find the center of gravity of the triangle.

10. Find the center of gravity of the area between the two curves $y = x^2$ and $y^2 = x$.

11. Find the center of gravity of the area between the curve $y = x^2$ and the line $y = x$.

12. Find the center of gravity of the area between the curve $y^2 = x$ and the line $y = x$.

13. Find the center of gravity of the area between the lines $y = x$ and $y = 2x$ from the origin to the line $x = 4$.

28. Mean Value. The average value of a set of quantities is the sum of the quantities divided by their number. For example, the average value of the record of a student whose tests were marked 25, 60, 60, 75, 100, 80, 40 per cent., would be

$$\frac{25 + 60 + 60 + 75 + 100 + 80 + 40}{7} = \frac{440}{7} = 62\frac{6}{7} \text{ per cent.}$$

The mean value of a set of quantities is the limit of the average value as the number of quantities is indefinitely increased. Thus, if $f(x)$ represent any one of the quantities and n the number of quantities, then

$$\text{Average value of } f(x) = \frac{\Sigma f(x)}{n} \qquad (49)$$

$$\text{Mean value of } f(x) = \lim_{n=\infty} \frac{\Sigma f(x)}{n} \qquad (50)$$

The form of (50) suggests an integral, but we can not pass to the limit and integrate without knowing with respect to what variable we integrate. The average value of $f(x)$ (49) can obviously be written:

$$\frac{\Sigma f(x) \, \Delta x}{n \, \Delta x}, \quad \frac{\Sigma f(x) \, \Delta y}{n \, \Delta y}, \text{ etc.}$$

and the mean value

$$\lim_{n=\infty} \frac{\Sigma f(x) \, \Delta x}{n \, \Delta x}, \quad \lim_{n=\infty} \frac{\Sigma f(x) \, \Delta y}{n \, \Delta y}, \text{ etc.}$$

or

$$\frac{\int_{x_1}^{x_2} f(x)\, dx}{\int_{x_1}^{x_2} dx},\quad \frac{\int_{y_1}^{y_2} f(x)\, dy}{\int_{y_1}^{y_2} dy},\ \text{etc.} \tag{51}$$

since, obviously $\displaystyle\lim_{n=\infty} n\,\Delta\,x = \int_{x_1}^{x_2} dx$, etc.

In the case when we multiply numerator and denominator by Δy and obtain the second form of (51) the relation between the variables x and y must be known or capable of determination.

Example. What is the mean value of the ordinate of the curve $y = x^2$, from the point (0, 0) to the point (2, 4)?

$$\text{M. V.} = \lim_{n=\infty} \frac{\Sigma\, x^2}{n}.$$

Shall we multiply numerator and denominator by Δx, Δy or by some other increment before integrating? The choice of the increment depends upon how we select the ordinates. If we space the ordinates equally along OX, we should multiply by Δx; if equally along OY, by Δy, etc. In the former case we should have

$$\text{M. V.} = \frac{\int_0^2 x^2\, dx}{\int_0^2 dx} = \frac{\dfrac{x^3}{3}\Big]_0^2}{x\Big]_0^2} = \tfrac{4}{3} = 1\tfrac{1}{3}$$

In the latter case

$$\text{M. V.} = \frac{\int_{y=0}^{y=4} x^2\, dy}{\int_0^4 dy} = \frac{\int_0^4 y\, dy}{\int_0^4 dy} = \frac{\dfrac{y^2}{2}\Big]_0^4}{y\Big]_0^4} = 2$$

We see that the mean value of the ordinate (or of any function) is dependent upon the manner in which values of the ordinate (or function) are chosen. Fig. 27 illustrates this in the case of the example just worked. It will be noted that spacing the points equally along OY crowds the ordinates together where they are longest.

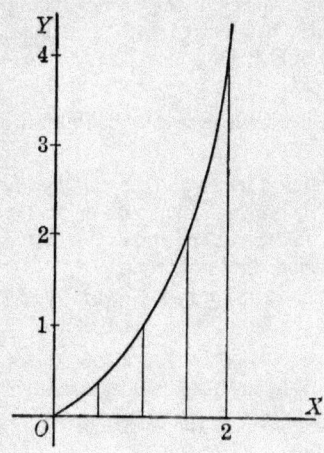 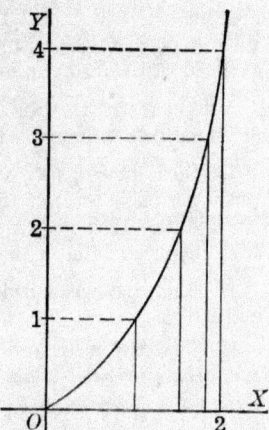

FIG. 27

EXAMPLES

1. Find the mean value of the ordinate of the curve $y = x^3$, from the point $(1, 1)$ to the point $(2, 8)$, (a) equally spaced along OX; (b) equally spaced along OY.

2. Find the mean value of the ordinate of an arch of the curve $y = \sin x$, spaced with respect to x.

3. A particle moves in a straight line according to the law $s = 5 + 6t - 3t^2$. Find the mean value of the velocity during the first second.

4. Find the mean value of the velocity of the particle in Example 3 between the distances 5 feet and 8 feet.

5. Find the mean distance from the origin at intervals of time during the first second, of the particle in Example 3.

6. A particle moves in a straight line according to the law $s = 4t - t^2$. Find the mean distance from the origin during the first two seconds; during the fifth second.

7. Find the mean value of the velocity of the particle in Example 6 during the first two seconds; during the fifth second.

8. A particle moves in a straight line with a velocity $v = 32t + 3$, starting at the origin. Find the mean velocity of the particle during the first three seconds. If the particle had moved with a constant velocity equal to the mean, how far would it have moved in the three seconds? How far did it actually move?

9. Find the mean value of the velocity with respect to the distance of the particle in Example 8.

10. Oil is poured into a tank containing water so that the density (mass per cubic unit) of the liquid is decreasing at the rate of 0.2 lbs. per minute. Find the density of the mixture at the end of 5 minutes. Find the mean value of the density during the 5 minutes.

11. Find the mean ordinate of $y = x^2 - 4x + 3$ between $x = 1$ and $x = 5$.

12. Find the mean ordinate of $y = x^2$ between $x = 0$ and $x = 3$, also between $x = -3$ and $x = +3$. Why are these results the same?

13. Find the mean value with respect to x of the square of the ordinate of a semicircle of radius a.

14. Find the mean sectional area of a sphere supposed cut by a series of equidistant parallel planes. Explain the result geometrically.

15. Find the mean sectional area of a cone of radius r and height h supposed cut by a series of planes parallel to the base.

16. If a body falls vertically from rest, its velocity v at the end of t seconds is given by the equation $v = 32t$. Find the average velocity, (a) for the first second, (b) for the first six seconds of its motion.

17. A quantity of steam expands so as to follow the law $pv^{0.8} = 200$, p being the pressure measured in lbs. per sq. in. Find the average pressure between the volumes 1 cu. ft. and 30 cu. ft.

18. A particle moves along the axis of x so that the force upon it at a distance x from the origin is equal to ax, where a is a constant. Find the mean value of the force as x increases from 0 to s.

19. If a body moves so that its speed is $v = t + \dfrac{1}{t^2}$, calculate the

distance traversed between the times $t = 2$ and $t = 4$ and the average speed.

20. Find the mean value of $\sin x$ between $x = \frac{\pi}{4}$ and $x = \frac{\pi}{3}$.

21. Find the mean value of $y = \dfrac{1}{x^2}$ between $x = 1$ and $x = 4$.

22. A spring oscillates so that the force of F lbs. which it exerts on a weight at the end of t seconds is given by $F = 2 \sin 3\, t$. Find the mean value of the force from $t = 0$ to $t = \frac{\pi}{3}$.

23. The electric current C in a conductor at time t is given by the equation $C = 4 \sin 200\, t$. Find the mean value of C from $t = 0$ to $t = \frac{\pi}{100}$ seconds.

29. Work, Attraction, Mass. The work done by (or against) a force is defined as the product of the force by the distance through which it acts. This definition can hold, however, only when the force is constant, unless we take the distance so short that the force can be considered as constant over that distance. We thus obtain the element of work, and by summing these elements, integrating, over the whole path we obtain the total work done.

Example 1. A coiled spring, two feet in length, resists compression with a force which varies as the distance through which the end of the spring is moved in compressing it. How much work is done in compressing the spring to half its original length?

Let x be the distance through which the end of the spring is compressed. Then the force, $F = K\,x$, and the work done in compressing it a very small distance, dx, is the element of work $d E = Kx\, dx$. Therefore,

$$E = \text{Work} = \int_0^1 Kx\, dx = \frac{K\,x^2}{2}\Big]_0^1 = \frac{K}{2}$$

The force of attraction between two masses is defined as the product of the masses divided by the square of the distance between them. This definition holds only when each mass is concentrated at a single point and the distance does not vary. If the mass is that of a body having dimensions, we must take a portion of the body so small that the whole portion can be considered to lie at a constant distance from the mass it is attracting. We thus form the element of attraction, the element of the thing to be summed, and by integrating we obtain the total force.

Example 2. Opposite the middle point of a very thin wire of length l cm. at a distance c centimeters from the wire, is situated a particle of unit mass. Find the force with which the wire attracts the particle. See Fig.

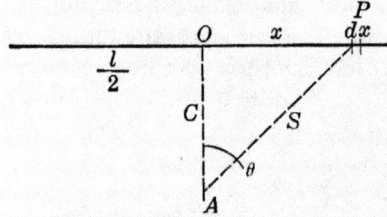

FIG. Ex. 2

Let x be the distance from the center of the wire to any point P. Let the density of the wire (mass per unit length) be ρ, a constant. Then the element of mass at P is $\rho\, dx$ and the element of force of attraction between $\rho\, dx$ and the particle at A is

$$d F = \frac{1 \cdot \rho\, dx}{S^2}$$

But this attraction is made up of two components one parallel and one perpendicular to the wire. The component parallel to the wire is neutralized by the pull of another little element, $\rho\, dx$, at the same distance from 0 as is P but on the other side, so that only the component perpendicular to the wire is effective. If we use OAP as the triangle of forces, AP will be the total force and AO the component at right angles to the wire; the normal component. Therefore we have for the normal attraction since $AO = AP \cos \theta$,

$$d F' = d F \cdot \cos \theta = \frac{\rho \cos \theta \, dx}{S^2}$$

and

$$F' = \int \frac{\rho \cos \theta \, dx}{S^2}.$$

This integral involves three variables, θ, x and S. Let us express it all in terms of the variable θ. Thus, since $x = c \tan \theta$, must $dx = c \sec^2 \theta \, d \theta$; and, since $S = c \sec \theta$, must $S^2 = c^2 \sec^2 \theta$. Therefore,

$$F' = \int_0^{\tan^{-1}\frac{l}{2c}} \frac{\rho \cos \theta}{c^2 \sec^2 \theta} \cdot c \sec^2 \theta \, d \theta$$

for the normal attraction of half the wire. It is obvious that θ begins with the value zero, and ends with the angle formed by joining the end of the wire to A. The tangent of this angle is $\dfrac{l}{\dfrac{2}{c}} = \dfrac{l}{2c}$, so that $\theta = \tan^{-1} \dfrac{l}{2c}$.

We then have for the total normal force,

$$2 F' = \frac{2\rho}{c} \int_0^{\tan^{-1}\frac{l}{2c}} \cos \theta \, d \theta = \frac{2\rho}{c} \left\{ \sin \theta \right\}_0^{\tan^{-1}\frac{l}{2c}}$$

$$= \frac{2\rho}{c} \left\{ \sin \left(\tan^{-1} \frac{l}{2c} \right) - \sin 0 \right\} = \frac{2\rho}{c} \cdot \frac{l}{\sqrt{l^2 + 4 c^2}}$$

$$= \frac{2 \rho l}{c \sqrt{l^2 + 4 c^2}}$$

If the total mass of the wire be M, then $\rho = \dfrac{M}{2l}$ and we may write

$$2 F' = \frac{M}{c \sqrt{l^2 + 4 c^2}}$$

Example 3. Suppose the density of the wire in Example 2 is not constant but equal to $\dfrac{l}{2} - x$, x being, as before, the distance from the center of the wire. What is the total mass of the wire?

In this case the element of mass, at P, is

$$dM = \rho \, dx = \left(\frac{l}{2} - x\right) dx$$

and

$$M = \int_{-\frac{l}{2}}^{+\frac{l}{2}} \left(\frac{l}{2} - x\right) dx = 2 \int_{0}^{\frac{l}{2}} \left(\frac{l}{2} - x\right) dx = \frac{l^2}{4}.$$

EXAMPLES

1. Under a circular metal plate of radius 20 centimeters is placed an electro-magnet which sets up a field of force in the plate such that the force, 500 dynes at the center, is always equal to $500 - r^2$ where r is the distance from the center. Find the work done in pushing a particle from the center to the edge of the plate.

2. A very thin wire of length 100 cm. and density 0.1 grammes attracts a unit particle in the line of the wire and 10 cm. from its end. Find the force of attraction between the wire and the particle.

3. A very thin homogeneous wire of length l and density ρ attracts a unit particle in the line of the wire and distant S units from its end. Find the force of attraction between the wire and the particle.

4. Force is defined as the product of mass by acceleration. A body of unit mass moving in a straight line has an acceleration of $+ K^2 x$ ft. per sec. per sec., where x is the distance from a fixed point 0 and K is a constant. Find the work done as the body moves from 0 to a distance C.

5. The weight of a body, which is a force, is inversely proportional to the square of its distance from the center of the earth. Find the work done in lifting a weight of 10 lbs. from the surface of the earth to a height of one mile above the surface. (Take the radius of the earth as 4000 miles).

6. Express as an integral the work done against attraction by moving

the particle of Example 2 from the end of the wire to its position 10 cm. away. (Use the result of Example 3.)

7. Express as an integral the work done in moving the particle of illustrative Example 2, Art. 29, from the middle point of the wire, 0, to the position A.

8. Find the total mass of a thin circular plate of radius a if the density at a distance x from the center is equal to $a - x$. (Suggestion: let the element of mass be that of a very narrow ring at any distance from the center.)

9. Find the mean value of the density of the plate of Example 8, with respect to distances from the center. Would a plate of constant density equal to this mean have a greater or less mass than the plate of Example 8?

10. A straight line bisects the angle between the axes of x and y and is 10 feet long. Find the mass of a triangular plate made by the given line, the axis of x and the line $x = 5\sqrt{2}$, (a) when the density varies as the distance from OX; (b) when the density varies as the distance from OY.

CHAPTER VIII

MISCELLANEOUS EXAMPLES

30. Miscellaneous Examples. Given the function $y = (x + 7)$ $(x - 2)(x - 4)$, answer questions 1–12.

1. Sketch the graph.

2. Plot the graph accurately from $x = -7$ to $x = 4$.

3. Find maximum and minimum values of the function.

4. Find the direction in which the graph crosses $OX; OY$.

5. Given that the abscissa increases at the rate of 2 feet per second, find the rate at which the ordinate is increasing at the points where the graph crosses $OX; OY$; at the maximum point; at the minimum point.

6. Find the mean value of the ordinate of the curve between $x = -7$ and $x = 2$; between $x = 2$ and $x = 4$; between $x = -7$ and $x = 4$.

7. Is the direction of the curve ever northeast; that is, at $45°$ with OX? Where?

8. Find the area between the curve and the axis of x from $x = -7$ to $x = 4$.

9. Find the abscissa of the center of gravity of the area between the curve and OX from $x = -7$ to $x = 2$.

10. Find the abscissa of the center of gravity of the area between the curve and OX from $x = 2$ to $x = 4$.

11. Find the abscissa of the center of gravity of the area bounded by the curve, the axis of x and the axis of y.

12. The abscissa of a point on the curve is measured as 6. Find the value of the corresponding ordinate and compute the approximate error in the ordinate due to a possible error of .02 in the abscissa.

13. Sketch the graphs of the four functions

(a) $y = x(x - 1)$
(b) $y = x(x - 1)^2$
(c) $y = x(x - 1)^3$
(d) $y = x(x - 1)^4$

126

What conclusion might be drawn from the behavior of the graphs at the point where $x = 1$?

14. Sketch the graph of the function $y = \dfrac{(x + 1)\,(x - 1)}{x + 2}$.

15. Sketch the graph of the function $y = \dfrac{(x - 1)\,(x - 3)}{x - 4}$.

16. If the tank of Ex. 10, Art. 28, is 10 ft. long, 4 ft. high and 5 ft. wide, find the pressure on the greatest side (a) at the beginning of the operation; (b) at the end of 5 minutes; (c) using the mean value of the density.

17. In Example 16 find the mean value of the pressure during the 5 minutes. Compare the result with result (c) of that example.

18. Find the mean value of the pressure in Example 16 with respect to the density.

19. A triangular plate, of negligible thickness, is of the form of the area between the coördinate axes and the line represented by $y = 3 - \dfrac{3\,x}{4}$. The density of any particle of the plate is equal to the distance of the particle from OY. Find the mass of the plate.

20. A tank whose length is 25 and height 10 is kept full by a stream of oil so that the density is decreasing at the rate of 2 units per minute. At what rate is the pressure on the face (25 by 10) changing?

21. In a triangle we have given the angle $B = 45°$, $C = 30°$, and the side $a = 100$ ft. What error is caused in the side c by an error in a of 0.2 ft., assuming B and C exact? (Suggestion: use the law of sines.)

22. Sketch the graph of the function
$$y = \frac{x + 1}{(x - 1)\,(x + 2)}$$

23. Sketch the graph of the function
$$y = \frac{x - 1}{(x - 3)\,(x - 4)}$$

24. A steamer has cast off and is drifting away from the pier with a velocity of 2 feet per second. A belated passenger jumps in a straight line from the dock (on a level with the steamer's deck) with a velocity of b feet per second, after the steamer has drifted 5 feet away. The wind, blowing directly from the steamer to the man, retards him with

a resistance of a feet per second per second. How long will the passenger take in making the leap?

25. In Example 24 if $a = 10$ ft. sec.^{2}, and $b = 12$ ft. sec., will the passenger make the steamer? In what time? What is the length of his leap?

26. In Example 24 if $a = 6$ ft. sec.^{2} and $b = 10$ ft. sec., in what time will the passenger make the steamer? With what length of leap? Explain your answers fully.

27. A particle is moving along the axis of x according to the law $x = 5 - 2t$, t being the time. Another particle moves along the axis of y according to the law $y = 4 - 3t$. How far apart are the particles when they start to move? When are they nearest to each other, and what is that distance? How long will they continue to approach?

28. A particle is moving on the axis of x according to the law $x = \cos t$; another particle on the axis of y according to the law $y = \sin t$. When are they nearest to each other? When are they moving with the same speed?

29. A particle is moving on OX according to the law $x = 2 \sin t$; another on OY according to the law $y = \sin 2t$. When are they nearest to each other? When farthest apart? What is the least distance? The greatest?

30. Find the area between the two arches above OX made by the curves $y = \sin x$ and $y = \dfrac{4x}{\pi} - \dfrac{4x^2}{\pi^2}$. Plot the two arches in one diagram on the same scale.

31. A parachute falls from a height of 365.3 feet, with an initial velocity of 3 feet per second, according to the law $a = 32 - 8t$. In what time will it again have a velocity of 3 ft. per sec.? How far will it then have dropped? What is its maximum velocity?

32. A belt transmits power, P, according to the law $P = V\left(T - \dfrac{WV^2}{g}\right)$ where V is the linear velocity of the belt in meters per second, W is the weight per meter in kilograms of the belt, and T is the tension in the belt while at rest. What value of V will make P a maximum? Find the maximum power when $W = 10$, $T = 2$. Note: $g = 9.81$ meters per sec. per sec.

33. If, in Example 32, the machinery is stopped so that V decreases at the rate of a meters per sec., what will be the rate of change of P?

34. In deep water the velocity of a wave of length l is given by the function $V = \sqrt{\dfrac{l}{a} + \dfrac{a}{l}}$, a being a known constant. What wave length will make the velocity a minimum?

35. The waste due to heat, depreciation, etc., in an electric conductor is given by the function $u = c^2 R + \dfrac{K^2}{R}$, where R is the resistance in ohms per mile, and c the current in amperes. If c is kept constant what value of R will make the waste least?

36. The gate at a grade crossing has an arm 15 feet long over the road and an arm 6 feet long over the sidewalk, the arms rotating upward on the same axis at the rate of 3 radians per minute. At what rate is the distance between the ends of the arms changing when they have rotated through 45° from the horizontal? Through 60°?

37. The rate at which the surface of a body of water falls is given by $\dfrac{dx}{dt} = \dfrac{s\sqrt{2\,g\,(a-x)}}{S}$ where a is the original depth of the water, s is the area of the cross-section of the opening through which the water is flowing, and S is the surface of the water. How long will it take to empty a vertical cylindrical tank 15 feet high, of radius 3 feet, through a hole of 1 inch radius in the bottom of the tank? How long will it take to lower the water 3 feet from a full tank? (Note: take $g = 32$ and t in seconds.)

38. The acceleration of a particle starting at a distance h from the center of the earth and falling toward the earth is given by the function $f = \dfrac{gR^2}{(h-S)^2}$, in which R is the radius of the earth, g the gravitational constant and S the distance through which the body has fallen. Find the velocity of the particle after it has fallen a distance S (a) from rest; (b) with initial velocity v_0. $\left(\text{Note:} f = \dfrac{dv}{dt} = \dfrac{dv}{ds} \cdot \dfrac{ds}{dt} = v\dfrac{dv}{ds}\right).$

39. In Example 38 find the velocity with which the particle reaches the earth, if it starts from rest (a) at a distance $2R$ from the center of the earth; (b) at an infinite distance. (Suggestion: In the result (a) of Example 38 put $s = h - R$.)

40. Taking $R = 4000$ miles and $g = 32$ feet sec.^{2}, compute the results of Example 39.

41. A curve is given by the equations $x = t^2$ and $y = 4t - t^3$, where t is the time in seconds, x and y being in feet. Sketch the curve. [Suggestion: Give values to t and plot the points (x, y)].

42. Find the area of the loop of the curve in Example 41. (Suggestion: Express the integral in terms of t.)

43. Find the volume got by revolving the loop of the curve in Example 41 about the axis of x. (Suggestion: Express the integral in terms of t.)

44. In what direction is the particle which generates the curve of example 41 moving at the end of 2 seconds?

45. At what angle does the curve of Example 41 cross the axis of x?

46. What is the speed of the particle in Example 41 when it starts to move? When it crosses the axis of x for the first time? At the highest point of the loop?

47. When and where will the components of velocity of the particle in Example 41 be equal?

48. On the ordinate of a point on the circle $x^2 + y^2 = a^2$, at right angles to the plane of the circle, a rectangle is constructed having the ordinate in question as base and the abscissa of the point as altitude. Find the mean value of the area of the rectangle as x passes from o to a.

49. Given a body moving according to the law $v = 32t$ (t in seconds), find (a) the distance traveled in the first $2\frac{1}{2}$ seconds; (b) the average value of the velocity during the time named.

50. Given a body moving according to the law $v = \sqrt{2gS}$, find the average value of the velocity during the first hundred feet. Compare the result with the result of Example 49.

51. Sketch the graphs of the three functions
  (a) $y = (x - 1)^2 (x + 2)$
  (b) $y = (x - 1)^3 (x + 2)$
  (c) $y = (x - 1)^4 (x + 2)$

52. Plot the graph of each of the functions in Example 51 at intervals $\Delta x = .1$ from $x = .5$ to $x = 1.5$.

53. Find the direction of each of the graphs of Example 51 at (a) $x = 1$; (b) $x = 1.1$.

54. Find the area under the arch for each curve in Example 51.

55. If each of the graphs in Example 51 is the path of a moving particle, and if the y-component of velocity in each of the three is 2 feet

per second, what is the speed in its path for each particle at the point whose abscissa is 1 foot? At the point whose abscissa is 1.1 foot?

56. Gravity acting downward, parallel to OY, find the turning moment with respect to OY, due to the force of gravity acting on a very thin plate in the shape of the first arch of the curve $y = \sin x$. The weight of any portion of the plate is proportional to the reciprocal of the distance of that portion from OY, and is equal to $\frac{1}{2}$ when $x = 2$.

57. The edges of a very thin plate in the form of an equilateral triangle have for their equations

$$y = \frac{4\sqrt{3}}{3} - x\sqrt{3}, \ y = \frac{4\sqrt{3}}{3} + x\sqrt{3}, \text{ and } y = -\frac{2\sqrt{3}}{3}.$$

A hole in the shape of an equilateral triangle with its sides parallel to the sides of the plate, is cut out of the center of the plate. The equations of the sides of the hole are $y = \frac{\sqrt{3}}{18} - x\sqrt{3}, \ y = \frac{\sqrt{3}}{18} + x\sqrt{3},$

and $y = -\frac{\sqrt{3}}{36}$. If the foot is the unit of length, show that the edges of the plate are 4 feet long; the edges of the hole 2 inches long. Also, if the weight of the material of the plate, in ounces per square foot, is equal to the distance from the vertex find the total weight of the plate. (Suggestion: Take strips of area parallel to OX.)

58. In example 19 find the mean value of the density of the plate. If the plate were of constant density equal to the mean what would be its mass?

59. Sketch the graph of the function $y = \sin x + 2 \sin \frac{x}{2}$. (Suggestion: put $y_1 = \sin x$, $y_2 = 2 \sin \frac{x}{2}$, plot these graphs in the same diagram and note that $y = y_1 + y_2$.)

60. Sketch the graph of the function $y = \cos x - \sin 2 x$.*

61. Sketch the graph of the function $y = \frac{1}{2} \sin x - \frac{1}{5} \sin 2 x$.*

62. Sketch the graph of the function $y = \frac{1}{2} \cos 2 x + \frac{1}{3} \cos 3 x$.*

63. Find maximum and minimum values of the functions given in Examples 59–62.

64. Find the area under the first arch of each of the curves given in Examples 59–61.

*See suggestion in Example 59.

CHAPTER IX

EXPONENTIAL AND LOGARITHMIC FUNCTIONS

31. Exponential Functions and their Graphs. An exponential function is one in which a variable occurs in the exponent. For example, $y = e^x$, $y = a^{\sin x}$ etc. The graphs of such functions can be plotted in the usual way by giving values to the independent variable (x), computing the corresponding values of the function (y), and plotting the points thus found.

Example 1. $y = e^x$.* See Fig. 28.

x	0	1	2	3	∞	-1	-2	-3	$-\infty$
y	1	e	e^2	e^3	e^∞	$\dfrac{1}{e}$	$\dfrac{1}{e^2}$	$\dfrac{1}{e^3}$	$\dfrac{1}{e^\infty}=0$
y	1	2.7	7.3	19.7	∞	.37	.14	.05	0

* The number e, the base of the Napierian or natural system of logarithms, may be defined as the limit of the sum of the terms of the following series as the number of terms increases without limit.

$$e = 1 + \frac{1}{1} + \frac{1}{\underline{|2}} + \frac{1}{\underline{|3}} + \frac{1}{\underline{|4}} + \ldots = 2.7182818 \ldots$$

In sketching graphs we shall use the approximate value $e = 2.7$.

Fig. 28

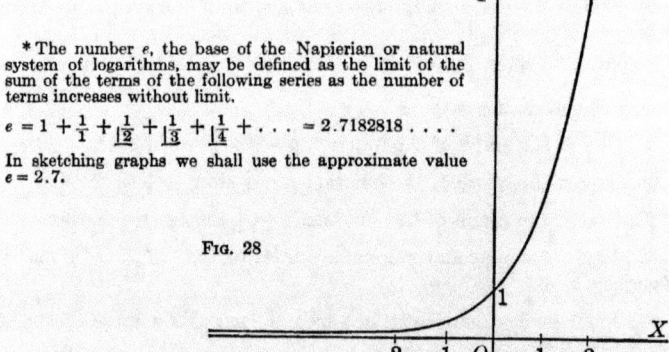

132

Example 2. $y = xe^{1-x}$

x	0	1	2	3	4	n	-1	-2	$-n$
y	0	1	$\dfrac{2}{e}$	$\dfrac{3}{e^2}$	$\dfrac{4}{e^3}$	$\dfrac{n}{e^{n-1}}$	$-e^2$	$-2\,e^3$	$-ne^{n+1}$
y	0	1	.74	.42	.20	0	-7.3	-39.4	$-\infty$

When $x = n$, and n is very large, e^{n-1} becomes very large even as compared with n, so that the fraction $\dfrac{n}{e^{n-1}}$, whose denominator is very large as compared with its numerator, becomes very small. We express this by writing, briefly, $y = \dfrac{\infty}{e^{\infty}} = 0$. Similarly, when $x = -n$ and n becomes, numerically, very large, we write, briefly, $y = -\infty \cdot e^{\infty} = -\infty$. See Fig. 29.

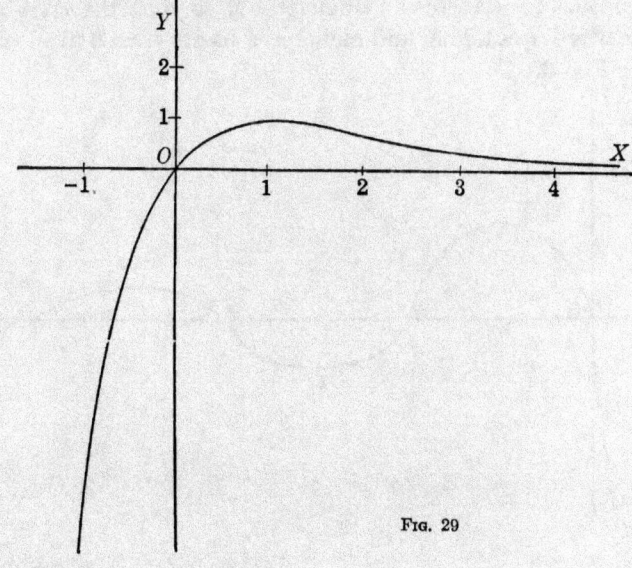

Fig. 29

Example 3. $y = e^{-x} \sin x$

Using, as in Art. 3, the limiting values of the sine we put

Sin $x = 0$ Sin $x = 1$

$x = 0, \pi, 2\pi \ldots - \pi, -2\pi$ $x = \dfrac{\pi}{2}, \dfrac{5\pi}{2} \ldots$

$y = 0, 0, 0 \ldots 0, 0$ $y = \dfrac{1}{e^{\frac{\pi}{2}}}, \dfrac{1}{e^{\frac{5\pi}{2}}} \ldots$

$$\text{Sin } x = -1$$

$$x = \frac{3\pi}{2}, \frac{7\pi}{2} \ldots$$

$$y = -\frac{1}{e^{\frac{3\pi}{2}}}, -\frac{1}{e^{\frac{7\pi}{2}}} \ldots$$

Thus the graph is a wave curve crossing the axis of x an infinite number of times at equal intervals (π), and with the crest of the wave lower and lower as x increases from zero. For negative values of x the factor e^{-x} is positive and increases as x increases numerically, so that the crest of the wave rises higher and higher as x passes from 0 to $-\infty$. See Fig. 30.

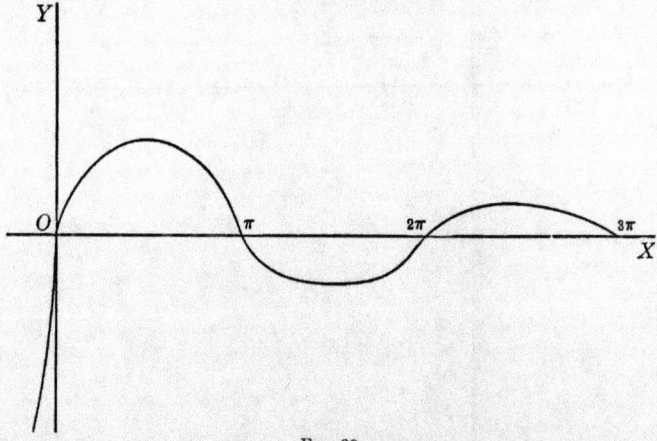

Fig. 30

32. Logarithmic Functions and their Graphs. A logarithmic function is one which involves the logarithm of a variable. For example, $y = \log_e x$, $y = \log_{10} \sqrt{1 - x^2}$, etc. We can plot such functions by giving values to the independent variable (x), computing or finding in tables, the corresponding values of the function (y) and plotting the points thus obtained. It is usually simpler, however, particularly in *sketching* the graphs of logarithmic functions, to treat them as the inverse of exponential functions and to proceed as in Art. 31.

Thus, given, Example 1, the function $y = \log_{10} x$,* we know, by the definition of a logarithm, that $x = 10^y$, which can be plotted by the methods of Art. 31. Thus (see Fig. 31).

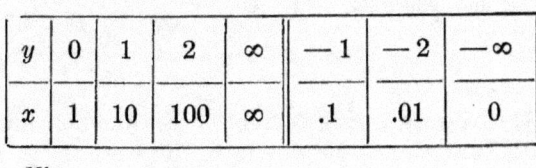

y	0	1	2	∞	-1	-2	$-∞$
x	1	10	100	∞	.1	.01	0

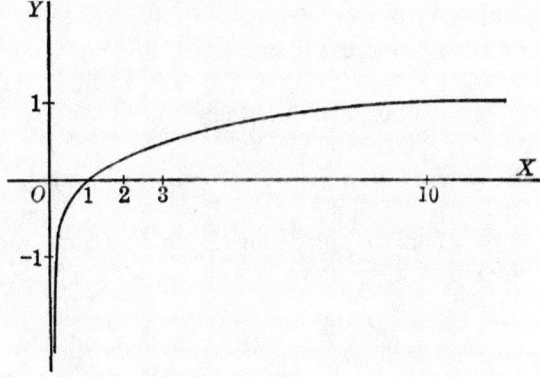

Fig. 31

* Hereafter when no base is mentioned the Napierian base, e, will be understood.

EXAMPLES

Sketch the graphs of the following functions:

1. $y = e^{-x^2}$ 2. $y = e^{1/x}$ 3. $y = x\, e^x$

4. $y = x\, e^{-x}$ 5. $y = x\, e^{1/x}$ 6. $y = x^2\, e^{-x}$

7. $y = e^{1-x}$ 8. $y = \frac{1}{2}\,(e^x + e^{-x})$ 9. $y = \frac{1}{2}\,(e^x - e^{-x})$

10. $y = e^{-x}\cos x$ 11. $y = e^{-x}\sin 2x$ 12. $y = e^{-2x}\cos x$

13. $y = \log_2 x$ 14. $y = \log_3 x$ 15. $y = \log_4 x$

16. $y = \log_{10}\sin x$ 17. $y = \log_{10}\cos x$ 18. $y = \log_{10}\tan x$

19. $y = \log\dfrac{1-x}{1+x}$ 20. $y = \log_{10}(x+1)$ 21. $y = \log(x+1)$

22. $y = \dfrac{a}{2}\left(e^{\frac{x}{a}} + e^{\frac{-x}{a}}\right)$ 23. $y = K\, e^{-h^2 x^2}$

24. Plot, in one diagram, the two curves

$$y = \tfrac{1}{2}\,(e^x + e^{-x}) \quad \text{and} \quad y - 1 = \frac{(e-1)^2}{2e}\, x^2$$

33. Derivatives and Integrals of Exponential and Logarithmic Functions.

We notice in the case of the second function of Art. 31 (Fig. 29) that it rises to some highest point and then descends; that the function has a maximum value. We have learned (Art. 15) how to find maximum and minimum values of a function by means of the derivative. Also in Chapters III to VII we have learned to apply the derivative and the integral to many problems connected with functions. All these applications can be made to exponential and logarithmic functions provided we can find their derivatives and integrals. This we shall now proceed to do.

Let $y = \log x$

Then $\Delta y = \log(x + \Delta x) - \log x$

$$= \log\frac{x + \Delta x}{x} = \log\left(1 + \frac{\Delta x}{x}\right)$$

and

$$\frac{\Delta y}{\Delta x} = \frac{\log\left(1 + \frac{\Delta x}{x}\right)}{\Delta x} = \log\left(1 + \frac{\Delta x}{x}\right)^{\frac{1}{\Delta x}}$$

which may be written

$$\frac{\Delta y}{\Delta x} = \frac{x \cdot \log\left(1 + \frac{\Delta x}{x}\right)^{\frac{1}{\Delta x}}}{x} = \frac{\log\left(1 + \frac{\Delta x}{x}\right)^{\frac{x}{\Delta x}}}{x}$$

Therefore

$$\underset{\Delta x = 0}{\text{limit}}\ \frac{\Delta y}{\Delta x} = \frac{\underset{\Delta x = 0}{\text{limit}}\ \log\left(1 + \frac{\Delta x}{x}\right)^{\frac{x}{\Delta x}}}{x}$$

$$= \frac{\log \underset{\Delta x = 0}{\text{limit}}\left(1 + \frac{\Delta x}{x}\right)^{\frac{x}{\Delta x}}}{x}$$

But $\quad \underset{}{\text{limit}}\left(1 + \frac{\Delta x}{x}\right)^{\frac{x}{\Delta x}} = e,^*$ and $\log e = 1$

Therefore

$$\frac{dy}{dx} = \frac{d \log x}{dx} = \frac{1}{x} \qquad\qquad (52)$$

* Let $\frac{\Delta x}{x} = h$, then $\frac{x}{\Delta x} = \frac{1}{h}$ and

$\underset{\Delta x = 0}{\text{limit}}\ \left(1 + \frac{\Delta x}{x}\right)^{\frac{x}{\Delta x}} = \underset{h = 0}{\text{limit}}\ (1 + h)^{\frac{1}{h}}$ = [expanding by the binomial theorem]

$\underset{h = 0}{\text{limit}}\left\{1 + \frac{1}{h} \cdot h + \frac{\frac{1}{h}\left(\frac{1}{h} - 1\right)}{\underline{|2}}\ h^2 + \frac{\frac{1}{h}\left(\frac{1}{h} - 1\right)\left(\frac{1}{h} - 2\right)}{\underline{|3}}\ h^3 + \ldots\right\}$

$= \underset{h = 0}{\text{limit}}\left\{1 + \frac{1}{1} + \frac{1 - h}{\underline{|2}} + \frac{(1 - h)(1 - 2h)}{\underline{|3}} + \ldots\right\}$

$= 1 + \frac{1}{1} + \frac{1}{\underline{|2}} + \frac{1}{\underline{|3}} + \ldots = e$

Remembering that integration is the reverse process of differentiation, it is obvious from (52) that

$$\int \frac{dx}{x} = \log x \qquad (53)$$

If $y = \log u$ we see, by foot-note Art. 13, page 41, and by (52) that

$$\frac{dy}{dx} = \frac{d \log u}{dx} = \frac{d \log u}{du} \cdot \frac{du}{dx} = \frac{1}{u} \cdot \frac{du}{dx} \qquad (54)$$

If $y = \log_a u$ we may write

$$y = \log_a u = \frac{\log_e u}{\log_e a} \, {}^*$$

Then

$$\frac{dy}{dx} = \frac{1}{\log a} \cdot \frac{d \log u}{dx} = \frac{1}{\log a} \cdot \frac{1}{u} \cdot \frac{du}{dx}$$

or

$$\frac{d \log_a u}{dx} = \frac{1}{u \log a} \cdot \frac{du}{dx} = \frac{\log_a e}{u} \cdot \frac{du}{dx} \qquad (55)$$

In particular

$$\frac{d \log_a x}{dx} = \frac{1}{x \log a} = \frac{\log_a e}{x} \qquad (56)$$

Consider next the function $y = e^u$

Then $\qquad u = \log y$

and

$$\frac{du}{dx} = \frac{d \log y}{dx} = \frac{d \log y}{dy} \cdot \frac{dy}{dx} = \frac{1}{y} \cdot \frac{dy}{dx}$$

or $\qquad \dfrac{dy}{dx} = y \dfrac{du}{dx}$

* Passano's Trigonometry, Art. 32.

That is $\quad \dfrac{de^u}{dx} = e^u \dfrac{du}{dx}, \quad$ and $\dfrac{de^x}{dx} = e^x.$ $\hfill (57)$

It is obvious that

$$\int e^x \, dx = e^x \qquad (58)$$

If $\hspace{4cm} y = a^u \hspace{3cm}$ we may write

$$\log y = u \log a$$

$$\frac{d \log y}{dx} = \log a \, \frac{du}{dx}$$

$$\frac{1}{y} \cdot \frac{dy}{dx} = \log a \, \frac{du}{dx}$$

$$\frac{dy}{dx} = y \log a \, \frac{du}{dx}$$

or $\hspace{3cm} \dfrac{da^u}{dx} = a^u \log a \, \dfrac{du}{dx} \hfill (59)$

Also $\hspace{3cm} \dfrac{da^x}{dx} = a^x \log a \hfill (60)$

It is obvious that

$$\int a^x \log a \, dx = \log a \int a^x \, dx = \int da^x = a^x$$

or $\hspace{3cm} \displaystyle\int a^x \, dx = \dfrac{a^x}{\log a} \hfill (61)$

For convenience of reference formulæ 52–61 are here assembled.

$$\frac{d \log u}{dx} = \frac{1}{u} \cdot \frac{du}{dx} \qquad\qquad \frac{d \log x}{dx} = \frac{1}{x}$$

$$\frac{d \log_a u}{dx} = \frac{1}{u \log a} \cdot \frac{du}{dx} \qquad \frac{d \log_a x}{dx} = \frac{1}{x \log a}$$

$$\frac{de^u}{dx} = e^u \frac{du}{dx} \qquad\qquad \frac{de^x}{dx} = e^x \qquad\qquad (62)$$

$$\frac{da^u}{dx} = a^u \log a \frac{du}{dx} \qquad\qquad \frac{da^x}{dx} = a^x \log a$$

$$\int \frac{dx}{x} = \log x, \quad \int e^x \, dx = e^x, \quad \int a^x \, dx = \frac{a^x}{\log a}$$

34. Integration by Parts. In addition to the three methods of integration given in Art. 18 there is a fourth, integration by parts, which is particularly useful in dealing with exponential and logarithmic functions.

We have proved, Art. 8, that

$$\frac{duv}{dx} = v \frac{du}{dx} + u \frac{dv}{dx}$$

or

$$\frac{duv}{dx} = vu' + uv'$$

where

$$u' = \frac{du}{dx} \quad \text{and} \quad v' = \frac{dv}{dx}$$

or

$$\int u' \, dx = u \quad \text{and} \quad \int v' \, dx = v$$

We may write, therefore,

$$d(uv) = vu' \, dx + uv' \, dx$$

or

$$vu' \, dx = d(uv) - uv' \, dx$$

Then

$$\int vu' \, dx = uv - \int uv' \, dx \qquad\qquad (63)$$

If then we have an integrand consisting of the product of two factors (vu') one of which (u') can be integrated separately, we can replace the integral of such an integrand by an expression of the form uv, minus a new integral which may be capable of integration by some of the processes already learned.

Formula (63) can best be learned in words.   Thus:

The integral of the product of two factors, one of which can be integrated separately, *equals* the product of the integral of that one factor, times the other; *minus* a new integral consisting of the same integral of the one factor times the derivative of the other.

The following mnemonic device may help in remembering the theorem:

$$\int \underset{u'}{(\text{one factor})} \underset{v}{(\text{other})} = \underset{u}{(\int \text{one})} \underset{v}{(\text{other})}$$

$$- \int \underset{u}{(\int \text{one})} \underset{v'}{(\text{derivative other})} \qquad (64)$$

Example 1.

$$\int xe^x dx = \int \underset{(\text{one})(\text{other})}{e^x \ x \ dx} = \underset{(\int\text{one})(\text{other})}{e^x \ . \ x} - \int e^x \ dx$$

$$= x \, e^x - e^x = e^x \, (x-1)$$

Example 2.

$$\int \log x \, dx = \int \underset{(\text{one})(\text{other})}{1 . \log x \, dx} = \underset{(\int \text{one})(\text{other})}{x \ \log x} - \int \underset{(\int\text{one})(\text{deriv. other})}{x \ . \ \frac{1}{x} \, dx}$$

$$= x \log x - x = x \, (\log x - 1)$$

Example 3.

$$\int x^3 \cos x \, dx = x^3 \sin x - 3 \int x^2 \sin x \, dx$$

(integrate again by parts)

$$= x^3 \sin x - 3 \left\{ - x^2 \cos x + 2 \int x \cos x \, dx \right\}$$

$$= x^3 \sin x + 3\,x^2 \cos x - 6 \left\{ x \sin x - \int \sin x\, dx \right\}$$

$$= x^3 \sin x + 3\,x^2 \cos x - 6\,x \sin x - 6 \cos x$$

$$= (x^3 - 6\,x) \sin x + (3\,x^2 - 6) \cos x$$

Example 4.

$$\int \cos^{-1} x\, dx = \int 1 \cdot \cos^{-1} x\, dx = x \cos^{-1} x + \int \frac{x\, dx}{\sqrt{1 - x^2}}$$

$$= x \cos^{-1} x - \sqrt{1 - x^2}$$

Example 5. Find maximum and minimum values of the function $y = xe^{1-x}$. (Ex. 2, Art. 31.)

$$\frac{dy}{dx} = e^{1-x} \frac{dx}{dx} + x \frac{de^{1-x}}{dx}$$

$$= e^{1-x} + x \cdot e^{1-x} \frac{d\,(1-x)}{dx}$$

$$= e^{1-x} - xe^{1-x} = e^{1-x}\,(1 - x) = 0$$

$$e^{1-x} = 0 \qquad 1 - x = 0$$

$$x = \infty \qquad\qquad x = 1$$

x	$=$	$\frac{1}{2}$	1	2
$\dfrac{dy}{dx}$	$=$	$+$	0	$-$
y	$=$	incr.	1	decr.
			Max.	

That is, the function has a maximum value, 1, when $x = 1$. This value would help us to sketch the graph in Fig. 29.

Example 6. At what angle does the curve $y = xe^{1-x}$ cross the axis of x? As in Example 5, $\dfrac{dy}{dx} = e^{1-x}\,(1 - x)$, which when $y = 0$ and, therefore, $x = 0$, has the value e. Therefore the curve crosses ox at $\tan^{-1} e = 69° 48'$.

Example 7. Find the area between the curve $y = xe^{1-x}$ from $x = 0$ to $x = \infty$.

$$A = \int_0^\infty xe^{1-x}\, dx = \int_0^\infty x \cdot e \cdot e^{-x}\, dx = e \int_0^\infty xe^{-x}\, dx$$

Integrating by parts

$$\int xe^{-x}\, dx = -xe^{-x} + \int e^{-x}\, dx = -xe^{-x} - e^{-x} = -\frac{(x+1)}{e^x}$$

Therefore

$$\text{Curve-area} = e\left[\frac{-(x+1)}{e^x}\right]_0^\infty = e\,(0+1) = e$$

Example 8. Find the mean value of $\log_{10} x$ between $x = 1$ and $x = 10$.

$$\text{M. V.} = \frac{\int_1^{10} \log_{10} x\, dx}{10-1}$$

Integrating by parts

$$\int 1 \cdot \log_{10} x\, dx = x \log_{10} x - \int x \cdot \frac{\log_{10} e}{x}\, dx$$

$$= x \log_{10} x - x \log_{10} e$$

Therefore

$$\text{M. V.} = \frac{x\,(\log_{10} x - \log_{10} e)\Big]_1^{10}}{9}$$

$$= \frac{10\,(1 - \log_{10} e) + \log_{10} e}{9}$$

$$= \frac{10 - 9\log_{10} e}{9} = 0.6768$$

Example 9. Find the rate of increase of the logarithm of a number as compared with the number, for the base e, the base 10, the base 3. When the number is greater than unity, does the logarithm increase faster or slower than the number?

Let $y = \log_a x$

$$\frac{dy}{dx} = \frac{1}{x \log a} = \frac{\log_a e}{x}$$

1°. Base e. $\dfrac{dy}{dx} = \dfrac{1}{x}$

2°. Base 10. $\dfrac{dy}{dx} = \dfrac{\log_{10} e}{x} = \dfrac{0.4343}{x}$

3°. Base 3. $\dfrac{dy}{dx} = \dfrac{1}{x \log 3} = \dfrac{1}{1.0986\, x}$

When $x > 1$ the logarithm increases slower than the number.

Example 10. $\displaystyle\int e^x \sin x \, dx$

Integrate by parts.

$$\int \underset{\text{(one)}}{e^x} \ \underset{\text{(other)}}{\sin x} \, dx = e^x \sin x - \int e^x \cos x \, dx \tag{a}$$

The new integral thus got is not simpler than the first. Let us integrate using $\sin x$ as the factor to be first integrated.

$$\int \underset{\text{(other)}}{e^x} \ \underset{\text{(one)}}{\sin x} \, dx = - e^x \cos x + \int e^x \cos x \, dx \tag{b}$$

We now notice that (a) and (b) are two equations which can be solved for $\displaystyle\int e^x \sin x \, dx$ by adding (a) and (b), and

for $\displaystyle\int e^x \cos x \, dx$ by subtracting (a) from (b). Thus

$$\int e^x \sin x \, dx = e^x \sin x - \int e^x \cos x \, dx$$

$$\int e^x \sin x \, dx = - e^x \cos x + \int e^x \cos x \, dx$$

Adding

$$2 \int e^x \sin x \, dx = e^x (\sin x - \cos x)$$

$$\int e^x \sin x \, dx = \frac{e^x (\sin x - \cos x)}{2}$$

Subtracting

$$0 = - e^x (\sin x + \cos x) + 2 \int e^x \cos x \, dx$$

or

$$\int e^x \cos x \, dx = \frac{e^x (\sin x + \cos x)}{2}$$

35. The Compound Interest Law. This law, also known as the law of organic growth, and the simple mass law, states the fact that a variable quantity, y, has a rate of increase with respect to an independent variable, x, which is proportional to the quantity (y) itself.

Thus
$$\frac{dy}{dx} = Ky$$

where K is a constant.
We may write

$$\frac{dy}{y} = K \, dx \quad \text{and} \quad \int \frac{dy}{y} = K \int dx$$

whence

$$\log y = Kx + c_1$$

$$y = e^{Kx + c_1} = e^{Kx} \cdot e^{c_1} = c e^{Kx}$$

and we see that the function or quantity whose rate of change is proportional to itself is the exponential function ce^{kx}.

Example 1. How much would \$100 amount to in 10 years, at 6 per cent per annum, if the interest were compounded each instant? *

Here

$$\frac{dy}{dt} = .06\, y; \quad \frac{dy}{y} = .06\, dt \qquad \text{and}$$

$$\log y = .06\, t + c \quad \text{or} \quad y = ce^{.06\, t}$$

But when $\quad t = 0,\ y = \$100 \quad \therefore\ c = 100 \quad$ and

$$y = 100\, e^{.06\, t}$$

When $\quad t = 10 \quad$ this becomes

$$y = 100\, e^{.6} = \$182.20$$

Example 2. Newton's law of cooling for an object cooled in moving air or running water is given by $\dfrac{d\theta}{dt} = -\, K\theta$ where t is the time and θ the difference in temperature between the object and the fluid. An object at 100° centigrade cools for 1 hour according to the law $\dfrac{d\theta}{dt} = -\, .5\,\theta$. What is its temperature at the end of the hour?

$$\frac{d\theta}{dt} = -\,0.5\,\theta; \quad \text{whence} \quad \theta = ce^{-0.5t}$$

when $t = 0,\ \theta = 100 \quad \therefore\ c = 100$ and $\ \theta = 100\, e^{-0.5t}$.

When $t = 1,\quad \theta = 60.6°$.

---

* Of course in actual practice interest is compounded at some definite period such as 6 months or a year.

How long will it take the body to cool from $100°$ to $10°$?

$$\theta = 100\,e^{-0.5t}, \quad \theta = 10 \text{ gives } 10 = 100\,e^{-0.5t} \quad \text{or}$$

$$e^{0.5t} = 10 \quad \therefore\ 0.5\,t\,\log_{10} e = \log_{10} 10 = 1$$

and
$$t = \frac{1}{0.5\,\log_{10} e} = 4 \text{ hrs. } 36 \text{ min.}$$

Or we might proceed thus

$$\int_0^t dt = \int_{100}^{10} \frac{d\theta}{-0.5\,\theta}$$

$$t = -2\log\theta\,\Big]_{100}^{10} = -2\,(\log 10 - \log 100) = 2\log 10 = 4.606$$

$$= 4 \text{ hrs. } 36 \text{ min.}$$

EXAMPLES

Find maximum and minimum values of the following functions:

1. $y = e^{-x}$ 2. $y = e^{\frac{1}{x}}$ 3. $y = xe^{x}$

4. $y = xe^{-x}$ 5. $y = xe^{\frac{1}{x}}$ 6. $y = x^2 e^{-x}$

7. $y = e^{1-x}$ 8. $y = \frac{1}{2}\,(e^x + e^{-x})$ 9. $y = \frac{1}{2}\,(e^x - e^{-x})$

10. Find the area under the first and second complete arches of the curve

$$y = e^{-x} \cos x.$$

11. Find the area under the first and second complete arches of the curve $y = e^{-x} \sin 2\,x$.

12. Find points at which the curves $y = e^{-x} \sin x$ and $y = e^{-x}$ have the same slope.

13. Find points at which the curves $y = e^{-x} \cos x$ and $y = e^{-x}$ have the same slope.

In the following examples x is the distance in feet, t the time in seconds, of a particle moving in a straight line according to the law given by the function. Find the position, velocity and acceleration of the particle at the time named; when the velocity will be zero; the maximum velocity.

14. $x = e^{-\frac{t}{10}} \sin 2\,t;\quad t = 5.$

15. $x = 5\,e^{-t} \cos t;\quad t = 1.$

16. $x = 2\,e^{\frac{t}{2}} \sin 2\,t;\quad t = \frac{\pi}{4}$

17. $x = 4\,e^{-t} \sin t;\quad t = \frac{\pi}{6}$

Find $\dfrac{dy}{dx}$ for the following functions:

18. $y = \log \dfrac{1-x}{1+x}$ 　　　　　19. $y = \log \dfrac{1+x}{1-x}$

20. $y = \log \dfrac{x}{x-1}$ 　　　　　21. $y = \log \dfrac{x}{x+1}$

22. $y = \log \left\{ x\,(x+1) \right\}$ 　　　23. $y = \log \sqrt{x^2 + 2\,x}$

24. Find the area between the axis of x and the curve $xy = 5$ from $x = 1$ to $x = 5$.

25. The relation connecting the pressure and volume of a certain gas is expressed by $pv^{1\cdot2} = 8$. Sketch the graph of the function and find (a) the area between the curve and the axis of p from $p = 1$ to $p = 4$; (b) the area between the curve and the axis of v from $v = 1$ to $v = 4$.

Integrate the following expressions:

26. $\displaystyle\int 2^x\,dx$ 　　　27. $\displaystyle\int 5^{2x}\,dx$ 　　　28. $\displaystyle\int \pi^{4x}\,dx$

29. $\displaystyle\int e^{2x+1}\,dx$ 　　30. $\displaystyle\int \frac{e^{\log 4\,x}}{x}\,dx$ 　31. $\displaystyle\int \frac{\sin x}{e^{\log \cos x}}\,dx$

32. $\displaystyle\int \frac{dx}{2\,x+1}$ 　　33. $\displaystyle\int \frac{dx}{x-5}$ 　　34. $\displaystyle\int \frac{dx}{3\,x+2}$

35. $\displaystyle\int \frac{e^x\,dx}{e^x+1}$ 　　36. $\displaystyle\int \frac{e^{2x}\,dx}{2\,e^{2x}-5}$ 　37. $\displaystyle\int \frac{dx}{x \log x}$

38. $\displaystyle\int \frac{\cos x\,dx}{1+\sin x}$ 　39. $\displaystyle\int \frac{\sec^2 x\,dx}{2\tan x+3}$ 　40. $\displaystyle\int \frac{\sin 2\,x\,dx}{\cos 2\,x+4}$

41. Find the area between the axis of x and the curve $y = e^x - e^{-x}$ from $x = 0$ to $x = 2$.

42. Find the area between the axis of x and the curve $y = e^{-\frac{x}{4}}$, $\sin x$ from $x = 0$ to $x = \pi$.

Integrate the following expressions:

43. $\int x e^{2x}\, dx$

44. $\int x \sin x\, dx$

45. $\int x \cos 2x\, dx$

46. $\int x^2 \sin x\, dx$

47. $\int \sin^{-1} x\, dx$

48. $\int \tan^{-1} x\, dx$

49. $\int e^x \sin e^x dx$

50. $\int x e^{x^2}\, dx$

51. $\int x^6\, dx.$

Find the volume got by revolving about the axis of x each of the following curves:

52. $y = e^x$ from $x = 0$ to $x = 2$.

53. $y = e^{-x}$ from $x = 0$ to $x = \infty$.

54. $y = e^x + e^{-x}$ from $x = -1$ to $x = 1$.

55. $y = x^{1/2} e^x$ from $x = 0$ to $x = 1$.

56. A particle moves in a straight line according to the law $s = a(e^t + e^{-t})$. Find (a) the velocity and acceleration at any time t; (b) the minimum velocity.

57. The acceleration of a particle moving in a straight line is given by $\dfrac{d^2s}{dt^2} = a(e^t + e^{-t})$. If it starts from rest at $t = 0$, how far will it travel in the first 5 seconds? What will be its velocity at the end of the 5 seconds?

58. The Napierian logarithm of the number 6.70 is calculated as 1.90209. If there is an error of 0.0002 in the logarithm, what error is caused in the number?

59. The number corresponding to a known Napierian logarithm, 2.21047, is calculated as 9.11. If there is an error of 0.01 in the number what error is caused in the logarithm?

60. The common logarithm of the number 2564 is calculated as 3.40890. If there is an error of .00002 in the logarithm what error is caused in the number?

61. The increase per second in the number of bacteria in a cubic inch of culture is proportional to the number of bacteria present. If N is the number of thousand bacteria per cubic inch and t is the time in seconds, express N as a function of t.

62. In Example 61 how long would it take the bacteria to increase from N_1 to N_2 per cubic inch?

63. The rate of decrease, with respect to the height, of atmospheric pressure above the earth's surface is proportional to the pressure. Express the pressure p as a function of the height, h.

64. In Example 63 if the pressure is 762 mm. at sea level, express p as a function of h; also if $p = 30$ in. at sea level.

65. A body cools in moving air according to Newton's law $\dfrac{d\theta}{dt} = -K\theta$. If θ falls from 40° C. to 30° C. in 200 seconds what is the value of the constant K? How long will it take for θ to fall from 30° to 20°? From 20° to 10°?

66. If a rotating wheel is stopped by water friction the rate of decrease of angular speed is proportional to the speed. $\left(\dfrac{d\omega}{dt} = -K\omega\right)$. Express the angular speed as a function of the time. Find the value of K if the speed of the wheel decreases 50 per cent in one minute.

67. How long will it take the wheel of Example 66 to slow down from 100 revolutions per minute to 10 revolutions per minute?

68. Find the mean value of the ordinate of the curve $y = e^x$ from $x = 0$ to $x = 1$.

69. Find the mean value of $\log x$ from $x = 1$ to $x = 10$.

70. Find the mean value of $\log_{10} x$ from $x = 10$ to $x = 100$.

71. A body moves so that its velocity (in feet per second) is proportional to the distance travelled. Express the distance as a function of the time.

72. If the body of Example 71 travels 100 feet in 10 seconds and 200 feet in 15 seconds, how far will it travel in t seconds?

73. The acceleration of a body moving in a straight line is proportional to the distance travelled. If the velocity is zero when the distance is zero express the distance, s, as a function of t, the time. $\left(\text{Suggestion: } \dfrac{d^2s}{dt^2} = \dfrac{dv}{ds} \cdot \dfrac{ds}{dt} = v\dfrac{dv}{ds}\right).$

74. The acceleration of a body moving in a straight line is inversely proportional to the distance travelled. The acceleration is 1 ft. sec.^{2} when the distance is 1 ft. What is the velocity of the body when it has travelled 10 feet, if $v = 0$ when $s = 1$?

75. A centre of force at a point, 0, attracts a particle, which is at rest a feet from 0, so that the acceleration of the particle is equal at any instant to its distance from 0. Find the velocity of the particle when it has travelled halfway to 0. (See suggestion under Example 73.)

ANSWERS

CHAPTER I

Art. 1

1. 8 cu. ft.; 343 cu. ft.; 29.8 cu. cm.

2. $\dfrac{s\sqrt{\pi s}}{6\pi}$

3. $v = s^3; v = \dfrac{h^3}{27}$

4. $I = \dfrac{kc}{100}; I = \dfrac{50\,k}{d^2}$; the same, $\dfrac{k}{200}$

5. 3; 24; 21

6. 3; 6; 3

7. $2\frac{1}{2}; 8\frac{1}{8}; 5\frac{5}{8}$

8. 0; 0.6021

9. 1.5574; — 0.8637; — 2.4211

10. 2; 16; 14

11. — 2; — 44; — 42

12. 0; — 60

13. — 4.8; — 19.75; — 14.95

14. 0; — 12

Art. 3

17. Starts from $(0, -2)$; at end 10th sec. $(20, 8)$

18. $(3, 3)$

19. $(3.8, -0.2)$

20. $(1, \frac{1}{2})$

21. $(\pm 2.4, \pm 2.4)$

22. $(2.28, 3.28); (-3.28, -2.28)$

23. $(0, 0), (\pm .5, \pm .5)$

24. $(.73, 2)$ etc.

25. $(0, 1), (6.28, 1)$ etc.

26. $(1.05, 1.73)$ etc.

27. $4\frac{1}{2}$ sec., $8\frac{1}{2}$ ft.

28. $3.5 +$ sec., $7.5 +$ ft.

29. 2.9 sec.; 6.9 ft. and $3.4 +$ ft.

30. 4.7 sec.; 8.7 ft. and 17.4 ft.

CHAPTER II

Art. 5

1. $4x$

2. $\dfrac{x}{2} + 1$

3. $6x - 2$

4. $3t^2$

5. $3v^2 - \dfrac{1}{v^2}$

6. $2t + \dfrac{2}{t^3}$

7. $\underset{\Delta x = 0}{\text{limit}} \dfrac{\Delta y}{\Delta x} = -\dfrac{1}{x^2}$;

$\underset{\Delta y = 0}{\text{limit}} \dfrac{\Delta x}{\Delta y} = -\dfrac{1}{(y-1)^2}$

8. $\displaystyle\lim_{\Delta x = 0} \frac{\Delta y}{\Delta x} = 1 - \frac{6}{x^7}$

9. $\displaystyle\lim_{\Delta x = 0} \frac{\Delta y}{\Delta x} = \frac{1}{2\sqrt{x}};$

$\displaystyle\lim_{\Delta y = 0} \frac{\Delta x}{\Delta y} = 2\,y$

10. $\displaystyle\lim_{\Delta x = 0} \frac{\Delta y}{\Delta x} = -\frac{3}{(x-2)^2};$ $\displaystyle\lim_{\Delta y = 0} \frac{\Delta x}{\Delta y} = -\frac{3}{(y-1)^2}$

11. $\displaystyle\lim_{\Delta x = 0} \frac{\Delta y}{\Delta x} = -\frac{4}{(3+x)^2};$ $\displaystyle\lim_{\Delta y = 0} \frac{\Delta x}{\Delta y} = -\frac{4}{(y+1)^2}$

12. $\displaystyle\lim_{\Delta v = 0} \frac{\Delta p}{\Delta v} = -\frac{a}{v^2};$ $\displaystyle\lim_{\Delta p = 0} \frac{\Delta v}{\Delta p} = -\frac{a}{p^2}$

Art. 10

1. $9\,x^2 - 2$

2. $2\,x - 4\,x^3 + 3$

3. $3\,(x+3)^2$

4. $-\dfrac{2}{(2+x)^3}$

5. $6\,x\,(x^2 + 2)^2$

6. $-6\,x^2\,(3 - x^3)$

7. $\dfrac{3\,x^2 + 1}{3\,(x^3 + x)^{2/3}}$

8. $\dfrac{1 - 4\,x}{2\sqrt{x - 2\,x^2}}$

9. $\cos 2\,x$

10. $\sin x\,(1 + \sec^2 x)$ **11.** $2\sec 2\,x\,\tan 2\,x - \sin x$

12. $2\cos 2\,x - 3\sec^2 3\,x$ **13.** $-\dfrac{2\,(u^2 + 4)}{(u^2 - 4)^2}$

14. $\dfrac{s^2 - 28\,s - 98}{s^2\,(s+7)^2}$ **15.** $-\dfrac{5\,(3\,y^2 + 2)}{y^2\,(y^2 + 2)^2}$ **16.** $\dfrac{3\,(4\,y^3 - 1)}{y^2\,(1 - y^3)^2}$

17. $8.1\,x^{1.7}$ **18.** $-\dfrac{25.1}{x^{6.02}}$ **19.** $1.57\,x^{2.14}$

20. $\dfrac{9}{2\,x^{13/4}}$ **21.** $\dfrac{dy}{dx} = -\dfrac{x}{y}$ **22.** $\dfrac{dy}{dx} = -\dfrac{4\,x}{5\,y}$

23. $\dfrac{dy}{dx} = \dfrac{4\,x}{5\,y}$ **24.** $\dfrac{dy}{dx} = -\dfrac{y}{x}$ **25.** $\dfrac{dy}{dx} = \dfrac{1 + y}{2 - x}$

26. $\dfrac{dy}{dx} = \dfrac{3 - y}{x - 1}$ **27.** $2\sin\left(4\,x + \frac{2\pi}{3}\right)$

28. $-3\cot^2\left(x - \frac{\pi}{4}\right)\csc^2\left(x - \frac{\pi}{4}\right)$

29. $\sec\left(2\,x + \frac{\pi}{6}\right)\left\{\,2\cos x\,\tan\left(2\,x + \frac{\pi}{6}\right) - \sin x\,\right\}$

30. $\cos x\,\cos 2\,x\,\tan 3\,x - 2\sin x\,\sin 2\,x\,\tan 3\,x + 3\sin x\,\cos 2\,x$
$$\sec^2 3\,x$$

CHAPTER III

Art. 13

1. $v = 6\,t - 3\,t^2$ ft. sec.
$f = 6 - 6\,t$ ft. sec.^{2}
Starts at $s = 2$ ft.
Moves E. 2 sec.; 6 ft.
Thereafter, W.

2. $v = 3\,t^2 - 6\,t$ ft. sec.
$f = 6\,t - 6$ ft. sec.^{2}
Starts at $s = 2$ ft.
Moves S. 2 sec.; 2 ft.
Thereafter, N.

3. $v = 18\,t - 5$ ft. sec.
$f = 18$ ft. sec.^{2}
Starts at $s = 7$ ft.
Moves $S.\,W.\,\frac{5}{18}$ sec.; $6\frac{11}{36}$ ft.
Thereafter, $N.\,E.$

4. $v = 4 - 2\,t$ ft. sec.
$f = -2$ ft. sec.^{2}
Starts at $s = 5$ ft.
Moves $N.\,W.$ 2 sec.; 9 ft.
Thereafter, $S.\,E.$

5. $v = 3\,t^2 - 6\,t$ ft. sec.
$f = 6\,t - 6$ ft. sec.^{2}
Starts at $s = 0$
Moves W. 2 sec.; 4 ft.
Thereafter, E

6. $v = 6\,t - 3\,t^2$ ft. sec.
$f = 6 - 6\,t$ ft. sec.^{2}
Starts at $s = 0$
Moves N. 2 sec.; 4 ft.
Thereafter, S.

7. $v = 32\,t$, 64, 160 ft. sec.; $f = 32$ ft. sec.^{2}

8. $v = 32\,t + 3$, 99, 323 ft. sec.; $f = 32$ ft. sec.^{2}

9. 320 ft. sec.

10. 320.8 ft. sec.

11. 10 sec., — 1600 ft.

12. $6\frac{1}{4}$ sec., 625 ft.

13. 143.1 ft. sec.

14. 64 ft. sec.; 6.4 ft.; 6.56 ft.; .16 ft. = 2.4%.
.64 ft.; .6416 ft.; .0016 ft. = .25%.

15. $\frac{2}{3}$ in. per min.

16. .069 in. per sec.

17. $\frac{3}{\pi}$ in. per min.

18. $\frac{1}{\pi}$ cm. per sec.; faster.

19. $v = 2.65$; $10.22 - 14\,t$; — 17.78; neither — why?

20. $-\dfrac{6.7}{10^5}, \dfrac{2.67}{10^4}$

21. 1.009

22. 0.172

23. 305 rad. sec.,
120 rad. sec.^{2}

24. $200\,\pi - 40\,\pi\,t$, — $40\,\pi$,
1 rev. per sec., 5 min.

25. $1000\,\pi\,\sqrt{3}$ in. per min.,—$1000\,\pi$ in. per min.

26. — 4.2, 2.7, $t = 1$; 0, — 5, $t = 2\,\pi$.

27. 9, $\frac{9}{2}\sqrt{2}$, $\frac{9}{2}\sqrt{2}$.

28. Former by 0.024 sec.

CHAPTER IV
Art. 14

1. $4x$, 8, 8.2   2. $-6x$, 7, 6.7   3. $3x^2 + 2$, 14, 14.61

4. $3x^2 - 2x$, 21, 21.81   5. $-\dfrac{y}{x}$, $-\dfrac{5}{2}$, -2.38

6. $-\dfrac{40}{x^3}$, -5, -4.65 7. $3t^2 - 6t$, 9, 9.61

8. $6t - 3t^2$, -9, -9.61

9. Greater when $t = 2$; $v = 8$

10. Greater when $t = 2$; $v = 3\frac{3}{4}$

11. Greater when $t = 1$; $v = 12$

12. Greater when $t = 2$; $v = 12$

13. Greater when $t = 2$; $v = 12$

14. Greater when $t = 2$; $v = 3$

15. -2, 0, 1   16. 0, -2, $\sqrt{3}$ 17. 1, 2

18. $\frac{1}{2}$, 1   19. 0, ∞ 20. ∞, $-\frac{2}{3}\sqrt{3}$

Art. 15

1. Max. $\frac{32}{27}$ 2. Max. $\frac{4}{243}$ 3. Neither
   Min. 0   Min. 0   4. Neither
                                    5. Neither

6. Max. $\frac{16}{9}$ 7. Max. 1   8. Max. $\sqrt{2}$ 9. Max. $\sqrt{2}$
   Min. $-\frac{16}{9}$ Min. -1 Min. $-\sqrt{2}$ Min. $-\sqrt{2}$

10. Min. $2\frac{5}{6}$ ft. sec.   11. Max. $1\frac{1}{3}$ ft. sec.   12. Min. $14\frac{2}{3}$ ft. sec.

13. $2\frac{5}{2}$ ft., $2\frac{1}{2}$ ft.   14. 108.6 cu. in.

15. Height equals diameter of base.

16. $15406 +$ sq. yds.   17. Height $=$ width of sill $= 4.2$ ft.

18. $t = \dfrac{u \sin a}{g}$, $x = \dfrac{u^2 \sin^2 a}{2g}$

19. Height, $4\sqrt{3}$; diameter, $4\sqrt{6}$; volume, $96\,\pi\,\sqrt{3}$

20. Height, 8.6 ft.; diameter, 12.2 ft.; volume, 331 cu. ft.

21. $2\pi\left(1 - \sqrt{\tfrac{2}{3}}\right) = \dfrac{2\pi}{5}$, approximately.

22. Height, 11.5 ft.; surface, 52.6 sq. ft.

23. $\dfrac{16\pi}{9\sqrt{3}}$ cu. ft.

24. $a = 900,\quad b = 60000,\quad m = 3000 - \dfrac{2000}{\sqrt[3]{4}-1},\quad n = \dfrac{200\sqrt[3]{10}}{\sqrt[3]{4}-1},$

$V = 20\left(\dfrac{3\,b}{2\,n}\right)^{3/5}$

25. $\dfrac{\pi}{4}$ **26.** $\dfrac{a}{2}$ **27.** 3 hrs. 40 min.

28. No; nearest, $3\tfrac{1}{4}$ miles, after 25 minutes.

29. Stem, 12 ft. 6 in.; arm, 6 ft. 11 in., approx.

30. Length along boundary $= 1\tfrac{1}{3}$ times width.

31. Width (side of triangle), 8.2 ft.; height (of rectangle), 5.2 ft.

32. Diameter, $\dfrac{10}{\pi+1}$; side, $\dfrac{5}{2\,(\pi+1)}$

33. Radius, $5\sqrt[3]{\dfrac{30}{\pi}}$; height, $\dfrac{20}{3}\sqrt[3]{\dfrac{30}{\pi}}$ ft.

34. 6 ft. 10 in., approx.

Art. 16

1. $3x(x-2)\,dx$ **2.** $(1 + 2x - 4x^3)\,dx$ **3.** $\left(3x^2 - \tfrac{3}{2}x^{1/2}\right)dx$

4. $\dfrac{(3x^2+1)\,dx}{2\sqrt{x^3+x}}$ **5.** $\dfrac{2x\,dx}{3\,(x^2-2)^{2/3}}$ **6.** $\dfrac{4x\,dx}{3\,(x^2+1)^{1/3}}$

7. $3\sin 6x\,dx$ **8.** $5\sec^2 5x\,dx$ **9.** $\sec^2\dfrac{x}{2}\tan\dfrac{x}{2}\,dx$

10. $(1 + \cos x)\,dx$ **11.** $-3\csc^2\left(3x + \tfrac{\pi}{4}\right)dx$

12. $-2\sin\left(2x + \tfrac{\pi}{6}\right)dx$ **13.** $-\dfrac{y}{x}\,dx$ **14.** $-\dfrac{2y}{x}\,dx$

15. $-\dfrac{y}{2x}\,dx$ **16.** $-\dfrac{2x+y}{x+2y}\,dx$ **17.** $-\dfrac{x}{y}\,dx$

18. $\dfrac{2\,x}{3\,y}\,dx$

19. 115.2 π cu. ft.

20. 197 ft., 9.8 ft.

21. $\dfrac{25\,\pi}{4},\ \dfrac{9\,\pi}{4}$ cu. ft.

22. $dv = \pi\,r^2 \cot a\,dr$

23. 14.8 π sq. in., 136.9 π cu. in.

24. 0.16 sq. in.　　　　**25.** 9.6 cu. in.　　　　**26.** 24.24

27. In each case approximate value is unity.

Exact value (1) 1.00007525; (2) 1.00007475

28. .006

29. (a) $83\frac{1}{3}$ sq. ft.; (b) $62\frac{1}{2}$ sq. ft.; (c) — $145\frac{5}{6}$ sq. ft.

30. (a) 5 ft.; (b) — 5 ft.; (c) — 1.4 ft.

31. Errors in C and E both $\doteq$, $\mp$ .005; errors in C and E one $\doteq$, one $\mp$ $\mp$ .055; about .9%.

32. .1015　　　　**33.** .005 ft. sec.^{2}

34. .018; $\frac{1}{2}$ of 1%.　　　　**35.** .03; 4%.

36. — 0.025

37. (a) $\doteq$ .008; (b) $\mp$ .016; (c) — .008; (d) .024

CHAPTER V

Art. 18

1. $x + \frac{2}{3}\,x^{3/2}$

2. $\frac{2}{3}\,x^{3/2} + \dfrac{x^3}{3}$

3. $\dfrac{x^2}{2} - 2\sqrt{x}$

4. $\frac{3}{4}\,x^{2/3}\,(x^{2/3} - 2)$

5. $\dfrac{x^2}{2} + \frac{4}{5}\,x^{5/4}$

6. $\frac{2}{3}\,x^{3/2} + \frac{3}{4}\,x^{4/3}$

13. $x + \frac{4}{3}\,x^{3/2} + \dfrac{x^2}{2}$

14. $x - 3\,x^{2/3} + 3\,x^{1/3}$

15. $x + \dfrac{2}{x} - \dfrac{1}{3\,x^3}$

16. $\dfrac{z^5}{5} + 2\,z - \dfrac{3}{z^3}$

17. $\frac{3}{5}\,y^{5/3} - 2\,y + 3\,y^{1/3}$

18. $\dfrac{s^2}{2} + \frac{12}{11}\,s^{11/6} + \frac{3}{5}\,s^{5/3}$

19. $\frac{1}{30}(3-5x)^{-6}$

20. $-\dfrac{5}{2(1+2x)}$

21. $-\frac{2}{9}(4-3t)^{3/2}$

22. $\frac{1}{3}(2u+1)^{3/2}$

23. $\frac{1}{2}\sqrt{1+4\theta}$

24. $-\frac{2}{3}\sqrt{2-3t}$

25. $\frac{1}{8}(x^2+1)^4$

26. $-\frac{1}{9}(2-t^3)^3$

27. $\frac{1}{12}(3+u^3)^4$

28. $-\dfrac{1}{12(3+2x^3)^2}$

29. $\dfrac{1}{8(5-s^4)^2}$

30. $-\frac{1}{10}(2-x^5)^2$

31. $\frac{3}{5}\sqrt{2s^{5/3}+5}$

32. $-\frac{12}{7}\sqrt[3]{1-y^7/4}$

33. $\frac{4}{9}(2x^{3/2}+1)^{1/4}$

34. $\frac{1}{3}(x^2+2x)^{3/2}$

35. $\sqrt{x^2+2x}$

36. $\frac{1}{9}(3x^2-6x)^{3/2}$

37. $\frac{2}{3}\sin^3\dfrac{x}{2}$

38. $-\frac{1}{2}\csc 2u$

39. $-2\sqrt{\cos t}$

40. $-\frac{1}{12}\cos^4 3x$

41. $\frac{1}{6}\sec^2 3x$

42. $\frac{1}{8}\sin^2 4x$

43. $y=\frac{2}{3}x^3-\dfrac{x^2}{2}+5$

44. $y=\dfrac{x^2}{2}-\dfrac{x^4}{4}+\frac{7}{4}$

45. $y=\dfrac{x^2}{2}+\dfrac{2x^{3/2}}{3}-\frac{40}{3}$

46. $y=\dfrac{x^3}{3}+\dfrac{x^2}{2}+2x-\frac{11}{6}$

47. $y=\dfrac{5t^2}{2}+4t-\frac{63}{2}$

48. $y=4t-\dfrac{5t^2}{2}+6$

49. $y=16t^2-37t+26$

50. $y=16t^2-56t+44$

51. $y=2x^2+3x-3$

52. $y=x^2-5x+8$

53. $s=16t-2t^2-10$

54. $s=62+16t+\dfrac{3t^2}{2}$

55. $s=16t+\dfrac{13t^2}{12}-\frac{85}{12}$

56. $s=16t-\frac{3}{2}t^2-\frac{29}{2}$

57. $s=7.1t^2+c$; 28.4 ft., 21.3 ft., 35.5 ft.

58. $y=3x+c$

59. $y=\dfrac{3x^2}{2}-2x+c$

60. $y=3x-\dfrac{x^2}{2}-\dfrac{x^3}{3}+c$

61. $y=x^2+c$

62. $y=\dfrac{x^3}{3}-x^2+3x+c$

63. $y=x^4+c$

64. $s = 4t + c$

65. $s = t^2 + 3t + c$

66. $s = t^3 + t^2 - 3t + c$

67. $s = 4t - \dfrac{t^2}{2} + c$

68. $s = t^3 + c$

69. $s = -\dfrac{4}{t} + c$

CHAPTER VI
Art. 19

1. 15.12; 15.02   **2.** 15.25; 15.32   **3.** 10.5; 10.64

4. 35.75; 35.96   **5.** 1.81; 1.83   **6.** 2.62; 2.66

7. 1.69; 1.702   **8.** 3.97; 3.996   **9.** 41; 40.16

10. 61.5; 60.24   **11.** 1.396   **12.** 1.974

13. 1.396   **14.** 1.974   **15.** 119.25 sq. ft.

16. 1410 sq. ft.   **17.** 1260 sq. ft.   **18.** 5.32

19. 42.5   **20.** 1400 sq. ft.

CHAPTER VII
Art. 20

1. $\frac{2}{3}$ **2.** 63   **3.** $-51\frac{2}{3}$ **4.** 2   **5.** $-\frac{1}{2}$

6. $-\frac{1}{8}$ **7.** $\frac{1}{20}$ **8.** $7\frac{7}{12}$ **9.** $\dfrac{2\sqrt{2}-1}{3}$

Art. 22

1. $12\frac{2}{3}$ **2.** $-6\frac{2}{3}$ **3.** $10\frac{2}{3}$ **4.** 36   **5.** $1\frac{5}{6}$

6. $2\frac{2}{3}$ **7.** $1\frac{19}{27}$ **8.** 4   **9.** 40   **10.** 60

11. $\sqrt{2}$ **12.** 2   **13.** $\sqrt{2}$ **14.** 0 or 2

15. 1280 sq. ft.; $1422\frac{2}{3}$ sq. ft.

16. $5\frac{1}{3}$ **17.** $\sqrt{2}-1$ **18.** $2-\sqrt{2}$ **19.** $\frac{1}{2}$

20. $\sqrt{2}\left(1-\frac{\pi}{4}\right)$ **21.** $1+\frac{\sqrt{2}}{8}\,(\pi-4)$ **22.** $85\frac{1}{3}$ **23.** $\frac{1}{4}$

Art. 23

1. $\dfrac{136\,\pi}{3}$ **2.** $\dfrac{40\,\pi}{3}$ **3.** $\dfrac{5312\,\pi}{15}$ **4.** $\dfrac{235\,\pi}{6}$

5. $\dfrac{344\,\pi}{3}$ **6.** $\dfrac{8\,\pi}{3}\,(7+6\sqrt{3})$ **7.** $\frac{4}{3}\,\pi\,ab^2;\ \frac{4}{3}\,\pi\,a^2b$

8. $\frac{4}{3}\,\pi\,a^3$ **9.** $75\,\pi$ **10.** $\dfrac{16\,\pi}{15}$ **11.** $\dfrac{16\,\pi}{15}$ **12.** $36\,\pi$

13. $\dfrac{125}{3}\,(2-\sqrt{2})$ **14.** $\dfrac{\pi\,h^3}{3}$ **15.** $8\,\pi$ **16.** $\pi\left(1-\frac{\pi}{4}\right)$

17. $\dfrac{2\,\pi}{3}$ **18.** $\dfrac{\pi}{2}$

Art. 24

1. $3\frac{29}{32}$ tons   **2.** $7\frac{13}{16}$ tons   **3.** 6.6 tons

4. 13.2 tons   **5.** $\frac{25}{48}$ tons   **6.** $1\frac{1}{24}$ tons

7. 151.6 lbs.   **8.** $19\frac{13}{32}$ tons   **9.** 500 lbs.; $1\frac{1}{3}$ ft.

10. 8.4 in.   **11.** $\dfrac{a^3\sqrt{2}}{64}$ tons   **12.** $\frac{5}{12}$ tons

13. $\dfrac{64\sqrt{2}}{3}$ tons   **14.** $\dfrac{3\sqrt{3}}{5}$ tons   **15.** Side, $1\frac{1}{4}$ tons; end $\frac{3}{4}$ tons

16. $95156\frac{1}{4}$ tons

Art. 25

1. (a) $\dfrac{625\sqrt{3}}{64}$ ft. tons   (b) $\dfrac{1875\sqrt{3}}{64}$ ft. tons

(c) 38 ft. tons, nearly   (d) 134.9 ft. tons

(e) $1\frac{29}{96}$ ft. tons   (f) $3\frac{29}{32}$ ft. tons

2. $27\frac{33}{64}$ ft. tons   **3.** $30\frac{45}{64}$ ft. tons   **4.** $\frac{1}{2}$ ft. ton

Art. 26

1. (a) 112.5 ft.; (b) 71.5 ft.   **2.** 61 ft.

3. 5131 ft.   **5.** $\dfrac{4500}{\pi}$ revs.   **6.** $\dfrac{684}{\pi}$ revs.

7. $\frac{\pi}{2}$ sec.; 2 ft.   **8.** 4 ft.   **9.** $\frac{1}{8}$

10. 50 ft.; 5 sec.   **11.** $3\frac{1}{8}$ ft.; $1\frac{1}{4}$ sec.   **12.** $3\frac{1}{8}$ ft.; $1\frac{1}{4}$ sec.

13. 4 ft.; 4 sec.   **14.** $5\frac{1}{3}$ ft.; 4 sec.   **15.** $4\frac{1}{2}$ ft.; 3 sec.

16. $170\frac{2}{3}$ ft.; 16 sec.   **17.** $1\frac{1}{5}$ rads.; $\frac{3}{4}$ sec.; $2\frac{1}{4}$ ft.

18. $\frac{1}{8}$ rad.; $\frac{1}{2}$ sec.; 1 ft.   **19.** $2\frac{1}{4}$ rad.; $1\frac{1}{2}$ sec.; $4\frac{1}{2}$ ft.

20. $\frac{1}{24}$ rad.; $\frac{1}{2}$ sec.; 1 in.   **21.** $\frac{\pi^3}{24}$ rad.; $\frac{\pi}{2}$ sec.; $\frac{\pi^3}{12}$ ft.

22. $\frac{\pi^3}{96}$ rad.; $\frac{\pi}{2}$ sec.; $\frac{\pi^3}{48}$ ft.   **23.** 10 sec.; 250 ft.

24. $11\frac{2}{3}$ sec.; $5\frac{1}{7}$ ft. sec.²   **25.** 250 ft.

26. $\frac{1}{2}$ sec.; .15 sec.; .35 sec.; 700 ft. sec.

Art. 27

1. (a) $\dfrac{5\sqrt{3}}{2}$ ft.   (b) $\dfrac{15\sqrt{3}}{4}$ ft.   (c) 7.7 + ft. below surface.

     (d) 10.2 ft.   (e) $2\frac{1}{2}$ ft.   (f) $3\frac{3}{4}$ ft.

     (g) 1.4 ft. below top of gate.   (h) 2 ft.

2. $(4, 1\frac{1}{2})$ *   **3.** $(1\frac{1}{3}, 1)$ *   **4.** $(1, 1\frac{1}{3})$ *

5. $(2\frac{4}{15}, \frac{4}{5})$ * or $(2\frac{11}{15}, \frac{4}{5})$ **6.** $(2\frac{2}{5}, 0)$ **7.** $(0, 2\frac{2}{5})$

8. $(\frac{8}{5}, 1\frac{6}{7})$ **9.** $\left(0, \dfrac{a\sqrt{3}}{3}\right)$ **10.** $(\frac{9}{20}, \frac{9}{20})$

11. $(\frac{1}{2}, \frac{2}{5})$ **12.** $(\frac{2}{5}, \frac{1}{2})$ **13.** $(\frac{8}{3}, 4)$

Art. 28

1. $3\frac{3}{4}; 4\frac{1}{2}$ **2.** $\frac{2}{\pi}$ **3.** 3   **4.** 4   **5.** 7

6. $2\frac{2}{3}; 2\frac{1}{3}$ **7.** 2; 5   **8.** 51; 153; 153

9. 66.1 nearly   **10.** 61.5 lbs.; 62 lbs.   **11.** 2 or $1\frac{1}{3}$

12. 3; 3   **13.** $\dfrac{2a^2}{3}$ **14.** $\dfrac{2\pi a^2}{3}$ **15.** $\dfrac{\pi r^2}{3}$

16. 16 ft. sec.; 96 ft. sec.   **17.** 33.6 lbs. per sq. ft.

18. $\dfrac{as}{2}$ **19.** $6\frac{1}{4}; 3\frac{1}{8}$ **20.** 0.7911

21. $\frac{1}{4}$ **22.** $\frac{4}{\pi}$ **23.** 0

* One vertex at the origin.

Art. 29

1. $7333\frac{1}{3}$ ergs.

2. $\frac{1}{110}$ dyne.

3. $\frac{\rho l}{s\,(l+s)}$

4. $\frac{K^2 C^2}{2}$

5. 3.79 foot-tons

8. $\frac{\pi a^3}{3}$

9. $\frac{a}{2}$

10. (a) $\frac{125\sqrt{2}}{3}K$; (b) $\frac{250\sqrt{2}}{3}K$

CHAPTER VIII

Art. 30

3. 144.8 max.; -10.03 min.

4. Axis of x at $\tan^{-1} 99$, $\tan^{-1}(-18)$, $\tan^{-1} 22$; axis of y at $\tan^{-1}(-34)$.

5. Axis of x, 198, -36, 44 ft. sec.; axis of y, -68 ft. sec.; max. and min. pts., 0.

6. $87\frac{3}{4}$; $-6\frac{2}{3}$; 70.6 or 73.

7. Yes; at $x = -3.8$ and $x = 3.1$ **8.** 803.1

9. -3.1 **10.** 3.02 **11.** -3.38 or 0.63

12. 104; 1.72 **16.** (a) 2.5 tons; (b) 2.46 tons; (c) 2.48 tons

17. 2.48 tons   **18.** 2.48 tons   **19.** 8

20. 2500 units per min.   **21.** 0.1 ft.

24. $t = \dfrac{(b-2) \pm \sqrt{(b-2)^2 - 10\,a}}{a}$

25. Yes; 1 sec.; 7 ft.

26. 1 sec.; 7 ft.; or, (Query ?) $1\frac{2}{3}$ sec.; $8\frac{1}{3}$ ft.

27. (a) $\sqrt{41}$; (b) $1\frac{9}{13}$; $\dfrac{7\sqrt{13}}{13}$; (c) $1\frac{9}{13}$

28. Always same distance apart; $t = \dfrac{\pi}{4}, \dfrac{3\pi}{4}$, etc.

29. Nearest; $t = 0$, π, 2π etc.; 0

Farthest; $t = \dfrac{\pi}{2}, \dfrac{3\pi}{2}$ etc.; 2

30. $\dfrac{2\pi}{3} - 2$ **31.** 8 sec.; 365.3 ft.; 67 ft. sec.

32. $V = \sqrt{\dfrac{gT}{3W}}$; 1.08   **33.** $\left(T - \dfrac{3WV^2}{g}\right)a$

34. $l = a$ **35.** $R = \dfrac{K}{C}$ **36.** 0; — 13.2 ft. min.

37. 20 min. 55 sec.; 2 min. 13 sec.

38. $V = \sqrt{\dfrac{2\,g\,R^2 s}{h\,(h - s)}}$; $V = \sqrt{\dfrac{2\,g\,R^2 s}{h\,(h - s)} + v_0^2}$

39. $\sqrt{gR}$; $\sqrt{2\,gR}$ **40.** 4.9 and 7 miles per sec.

42. $17\frac{1}{15}$ sq. ft.   **43.** $\dfrac{64\pi}{3}$ cu. ft.   **44.** $\tan^{-1}(-2)$

45. $\frac{\pi}{2}$; $\tan^{-1}(-2)$ **46.** 4, 4, $\dfrac{4\sqrt{3}}{3}$ ft. sec.

47. $t = 0.9$ sec.; (0.8, 0.7)   **48.** $\dfrac{a^2}{3}$

49. 100 ft.; 40 ft. sec.   **50.** $53\frac{1}{3}$ ft. sec.

53. At $x = 1$, each is parallel to OX; at $x = 1.1$, 32° 13′, 5° 22′ and 0° 43′.

54. 6.75; — 12.15; 24.3

55. At $x = 1$ each is infinite

     At $x = 1.1$, 3.75 ft. sec., 21.37 ft. sec., 160.02 ft. sec.

56. 2   **57.** $12\frac{1}{2}$ oz. approx.   **58.** 2; 12

63. (a) Max. 2.598 at $x = \dfrac{2\pi}{3}$ etc.;

       Min. — 2.598 at $x = \dfrac{10\pi}{3}$ etc.

   (b) Max. 0.369 at $x = 122° 32′$
            1.760 at $x = 323° 37′$

       Min. — 0.369 at $x = 57° 28′$
            — 1.760 at $x = 216° 23′$

   (c) Max. 0.607 at $x = 117° 26′$
            Min. — 0.607 at $x = 242° 34′$

(d) Max. $\frac{5}{6}$ at $x = 0$

$\qquad$ 0.216 at $x = \dfrac{4\,\pi}{5}$

$\qquad$ Min. -0.674 at $x = \dfrac{2\,\pi}{5}$

$\qquad$ $\frac{1}{6}$ at $x = \pi$

64. (a) 8; (b) $-\frac{1}{4}$; (c) 1

CHAPTER IX
Art. 35

1. None $\qquad$ 2. None $\qquad$ 3. Min., $-\dfrac{1}{e}$ at $x = -1$

4. Max., $\dfrac{1}{e}$ at $x = 1$ $\qquad$ 5. Min., e at $x = 1$

6. Min., 0 at $x = 0$; Max. $\dfrac{4}{e^2}$ at $x = 2$

7. None $\qquad$ 8. Min., 1 at $x = 0$ $\qquad$ 9. None

10. $A_1 = -\dfrac{1 + e^{\pi}}{2\,e^{\frac{3}{2}}}$; $\quad A_2 = \dfrac{1 + e^{\pi}}{2\,e^{\frac{5\pi}{2}}}$

11. $A_1 = \frac{2}{5}\left\{1 + e^{-\frac{\pi}{2}}\right\}$; $A_2 = -\frac{2}{5}\left\{e^{-\frac{\pi}{2}} + e^{-\pi}\right\}$

12. $x = \dfrac{\pi}{2},\ \pi,\ \dfrac{5\,\pi}{2},\ 3\,\pi$ etc. $\qquad$ 13. $x = 0,\ \dfrac{\pi}{2},\ 2\,\pi,\ \dfrac{5\pi}{2}$ etc.

$\quad y = $ Qy.? $\qquad\qquad\qquad\qquad\quad y = $ Qy.?

14. $x = -\ .33$ ft. $\qquad\qquad$ 15. $x = .996$ ft.

$\quad v = -\ .99$ ft. sec. $\qquad\qquad\quad v = -2.54$ ft. sec.

$\quad f = 1.53$ ft. sec.^{2} $\qquad\qquad\quad f = 3.1$ ft. sec.^{2}

$\quad v = 0$ when $t = .76$ sec., $\qquad v = 0$ when $t = \dfrac{3\,\pi}{4},\ \dfrac{7\,\pi}{4}$

$\qquad$ 2.34 sec. etc. $\qquad\qquad\qquad\quad$ etc. sec.

$\quad$ Max. $v = 1.45$ ft. sec. $\qquad\qquad$ Max. $v = \dfrac{5}{e^{\pi}}$ ft. sec.; Max. v

$\qquad\qquad\qquad\qquad\qquad\qquad\qquad$ (numerical) 5 ft. sec.

16. $x = 2 e^{\frac{\pi}{8}}$ ft.

$v = e^{\frac{\pi}{8}}$ ft. sec.

$f = -\frac{15}{2} e^{\frac{\pi}{8}}$ ft. sec.²

$v = 0$ when $t = .9, 2.5$ etc. sec.

Max. $v = 4.5$ ft. sec.

17. $x = 2 e^{-\frac{\pi}{6}}$ ft.

$v = 1.46 e^{-\frac{\pi}{6}}$ ft. sec.

$f = -3.46 e^{-\frac{\pi}{6}}$ ft. sec.²

$v = 0$ when $t = \dfrac{\pi}{4}, \dfrac{5\pi}{4}$ etc. sec.

Max. $v = 4 e^{-\frac{3\pi}{2}}$; Max. v (numerical) $4 e^{-\frac{\pi}{2}}$

18. $\dfrac{2}{x^2 - 1}$ **19.** $\dfrac{2}{1 - x^2}$ **20.** $\dfrac{1}{x - x^2}$ **21.** $\dfrac{1}{x^2 + x}$

22. $\dfrac{2x + 1}{x^2 + x}$ **23.** $\dfrac{x + 1}{x^2 + 2x}$ **24.** 8.05

25. $A_1 = 8.82$; $A_2 = 9.68$ **26.** $\dfrac{2^x}{\log 2}$

27. $\dfrac{5^{2x}}{2 \log 5}$ **28.** $\dfrac{\pi^{4x}}{4 \log \pi}$ **29.** $\frac{1}{2} e^{2x+1}$

30. $4x$ **31.** $\log \sec x$ **32.** $\log \sqrt{2x + 1}$

33. $\log (x - 5)$ **34.** $\frac{1}{3} \log (3x + 2)$ **35.** $\log (e^x + 1)$

36. $\frac{1}{4} \log (2 e^{2x} - 5)$ **37.** $\log (\log x)$ **38.** $\log (1 + \sin x)$

39. $\log \sqrt{2 \tan x + 3}$ **40.** $-\frac{1}{2} \log (\cos 2x + 4)$ **41.** $\left(e - \dfrac{1}{e} \right)^2$

42. $\frac{16}{17} \left(\dfrac{1}{e^{\frac{1}{4}}} + 1 \right)$ **43.** $\dfrac{e^{2x}}{4} (2x - 1)$ **44.** $\sin x - x \cos x$

45. $\frac{1}{4} (2x \sin 2x + \cos 2x)$ **46.** $2 \cos x + 2x \sin x - x^2 \cos x$

47. $x \sin^{-1} x + \sqrt{1 - x^2}$ **48.** $x \tan^{-1} x - \log \sqrt{1 + x^2}$

49. $-\cos e^x$ **50.** $\frac{1}{2} e^{x^2}$ **51.** $\dfrac{x^{e+1}}{e + 1}$

52. $\frac{\pi}{2} (e^4 - 1)$ **53.** $\frac{\pi}{2}$ **54.** $\pi \left(e^2 + 2 - \dfrac{1}{e^2} \right)$

55. $\frac{\pi}{4} (e^2 + 1)$

56. $v = a (e^t - e^{-t})$; $f = a (e^t + e^{-t}) = s$; no minimum

57. $s = a \left(e^5 + e^{-5} - 2 \right); v = a \left(e^5 - e^{-5} \right)$

58. .00134   **59.** .0011   **60.** .1181

61. $N = CE^{kt}$ **62.** $\dfrac{1}{K} \log \dfrac{N_2}{N_1}$ **63.** $p = ce^{-kh}$

64. $762\, e^{-kh}; 30\, e^{-kh}$ **65.** $K = \dfrac{\log \frac{4}{3}}{200}$; 4 min. 42 sec.; 8 min. 2 sec.

66. $K = \log 2$ **67.** 3 min. 19 sec.   **68.** 1.718

69. 1.5584   **70.** 1.6768   **71.** $S = CE^{kt}$

72. $25\,(2)^{\frac{t}{5}}$ **73.** $S = CE^{\pm \sqrt{k}.t}$ **74.** 2.15 ft. sec.

75. $v = \pm \dfrac{a}{2} \sqrt{3}$